5th Target Fabrication Workshop 2014

Journal of Physics: Conference Series
Volume 713

St. Andrews, United Kingdom
6 – 11 July 2014

Editor:

Martin Tolley

ISBN: 978-1-5108-2452-2
ISSN: 1742-6588

Printed from e-media with permission by:

Curran Associates, Inc.
57 Morehouse Lane
Red Hook, NY 12571

Printed by Curran Associates, Inc. (2016)

Institute of Physics
Dirac House, Temple Back
Bristol BS1 6BE UK

Phone: 44 1 17 929 7481
Fax: 44 1 17 920 0979

techtracking@iop.org

Additional copies of this publication are available from:

Curran Associates, Inc.
57 Morehouse Lane
Red Hook, NY 12571 USA
Phone: 845-758-0400
Fax: 845-758-2633
Email: curran@proceedings.com
Web: www.proceedings.com

Table of contents

Volume 713

5th Target Fabrication Workshop
6–11 July 2014, St Andrews, UK

Accepted papers received: 18 April 2016
Published online: 24 May 2016

Preface

Papers

Journal of Physics: Conference Series **713** (2016) 011001 doi:10.1088/1742-6596/713/1/011001

5th Target Fabrication Workshop, St Andrews, Scotland

The fifth target fabrication workshop took place at St Andrews University, Scotland, 6-11 July 2014. Approximately 60 delegates took part, originating from the UK, USA, Japan, Spain, Italy, Russia, France and Germany. The workshop was on various subjects all relevant to the laser community and included: cryogenic targets, high repetition-rate targets, porous target materials, new techniques for the production of spherical targets, target manipulation and micro engineering, and spectroscopy of materials related to laser targets.

Attendees included representatives from institutions and companies such as Rutherford Appleton Laboratory, St Andrews University, University of Michigan, Lawrence Livermore National Laboratory, LULI, CEA, University of Darmstadt, Lebedev Physical Institute (Moscow), the Institute of Laser Engineering (Osaka), Tokyo Institute of Technology, Hamamatsu Corporation and Scitech Ltd.

Of particular significance was the format of the conference, which was conducted in presentation/workshop style, where delegates were encouraged to take part in questions and discussions after each presentation. Thus all the delegates had more chances to get involved in not just simple question-and-answer sessions, but also in real problem solving deliberations.

One other important aspect of this conference was that younger scientists were specifically encouraged to attend. As a result, more than 50% of the delegates were younger scientists and their contributions and the levels of complexity and quality of their research was outstanding. It is hoped that this will continue in the future and younger generations will contribute positively in this area of research, which is vitally important for the future of this field, or indeed any other field.

Among other subjects, several speakers addressed the topic of high repetition-rate targets needed for the future production of energy by means of Inertial Confinement Fusion (ICF). In this context, the low temperature laboratory of CEA in Grenoble developed a cryostat able to produce a continuous film of solid molecular hydrogen of some tens of microns in thickness and one millimetre in width. This type of target could be used to produce neutrons for ICF. Furthermore, representatives from Lebedev Physical Institute presented their ideas to use magnetic levitation as a means to manipulate cryogenic ICF targets, and another talk was about employing microfluidics to produce shells for the fuel pellets used in ICF.

In a survey conducted after the conference the feedback, not just for the workshop but also the venue and facilities, was very positive, particularly from the younger scientists. It was refreshing to note that younger scientists felt that their work was valued and considered important and they benefitted enormously from this workshop. Overall it was an extremely successful workshop.

One conclusion of the conference was that it would be useful to have more interaction between the designers of laser experiments and the materials people who are so heavily involved in realising the targets that make these experiments possible use. For that purpose, the experimentalists should attend this type of conference, so they can learn about the trends in target fabrication affecting their experiments.

This was the fifth target fabrication workshop and it was decided that if possible, this workshop should be organised again in two years' time with the same format encouraging younger scientists to attend. It goes without saying that all this would not have been possible without generous financial contributions from the sponsors. On behalf of all the delegates, I would like to thank all of them.

Finally, the weather was agreeable as well, which always helps!

Wigen Nazarov,
Chair of the local organising committee

5th Target Fabrication Workshop Conference Proceedings, Editor's Preface

The first Target Fabrication Workshop was held in 2006 in Abingdon, UK. In the intervening years activity in the field has expanded to develop new technologies and encourage young researchers. A lot of the work in Target Fabrication has been published in peer reviewed journals although, arguably, the discipline has still to find a natural home. Consequently it was decided to publish the Conference Proceedings as a standalone volume to both act as a record and also be a future resource.

Looking forward the discipline of Target Fabrication appears to be increasingly important for enabling high power/energy laser facilities to deliver cutting edge science. Furthermore with the shift to higher repetition rates one area of growth for laser systems is expected to be into beams and their applications for which, again, Target Fabrication will be key. Additionally experience to date with higher repetition rate laser systems has indicated that the challenge of high rate, high accuracy target positioning is inseparable from that of target production. Consequently a new discipline, which has been called Microtargetry, is beginning to emerge combining the microtarget production and positioning challenges. If Microtargetry grows as anticipated the necessary R&D will require a publication platform and it is hoped that these Conference Proceedings will act as an early indicator.

On behalf of my co-editors I would like to thank all who have contributed to these Proceedings.

Martin Tolley,
Principal Editor

Journal of Physics: Conference Series **713** (2016) 011003 doi:10.1088/1742-6596/713/1/011003

Peer review statement

All papers published in this volume of *Journal of Physics: Conference Series* have been peer reviewed through processes administered by the proceedings Editors. Reviews were conducted by expert referees to the professional and scientific standards expected of a proceedings journal published by IOP Publishing.

Journal of Physics: Conference Series **713** (2016) 012001 doi:10.1088/1742-6596/713/1/012001

Batch Production of Micron-scale Backlighter Targets

G. Arthur

Scitech Precision Ltd, Rutherford Appleton Laboratory, Harwell Oxford, Chilton, Oxon, OX11 0QX, UK

E-mail: graham.arthur@scitechprecision.com

Abstract. The fabrication of micron-scale backlighter targets is described. Traditionally laser targets have been fabricated using conventional machining or coarse etching processes and have been produced in quantities of 10s to low 100s. The processes described herein allow batch production with numbers in the 1000s. In addition, the Micro-Electro-Mechanical System (MEMS) fabrication techniques used allow much finer tolerances and more accurate placement of the various components relative to each other.

1. Introduction.

The generation of x-ray probe beams for diagnostic use in high power laser science is well known and has been in use for many years. It is commonly achieved by the irradiation of a high-Z metallic target positioned behind a pinhole and close to the primary target. These backlighter targets have had typical dimensions of several 10s of microns in diameter. As a consequence the source size from these is relatively large and hence the resolution of the x-ray probe beam is limited. There has, therefore, been a drive to fabricate backlighter targets with significantly smaller dimensions to create probe beams with greater resolution.

In addition, the pulse repetition rate of current and future high power laser systems is increasing and this, in turn, leads to an increase in the numbers of targets required. Traditionally laser targets, including backlighters, have been fabricated using conventional machining, coarse etching processes and hand assembly which allow targets to be produced in quantities of 10s to low 100s, but this will no longer be able to satisfy demand in both quantities and dimensions/tolerances for the future.

This paper will describe the mass production of micron-scale, micro-wire backlighter targets using Micro-Electro-Mechanical System (MEMS) fabrication techniques. These techniques are a development of the fabrication methods using in the semiconductor industry and use industry standard silicon wafers as the basic substrate. In the same way that integrated circuits (ICs) are manufactured in very large numbers, it should be possible to batch-produce micro-wire backlighter targets on silicon wafers with numbers in the 1000s, even for short production runs.

2. Backlighter target design.

2.1. Conventional backlighter targets.
A conventional backlighter target design typically consists of a high-Z metallic source behind a pinhole. The example shown in Figure 1 is a design used on the Orion laser and consists of a 400μm diameter x 0.5μm thick titanium micro-dot on a 2mm square x 50μm thick CH film. This is mounted 0.5mm behind a 5mm square x 50μm thick tantalum pinhole which has a 10μm CH coating on all surfaces.

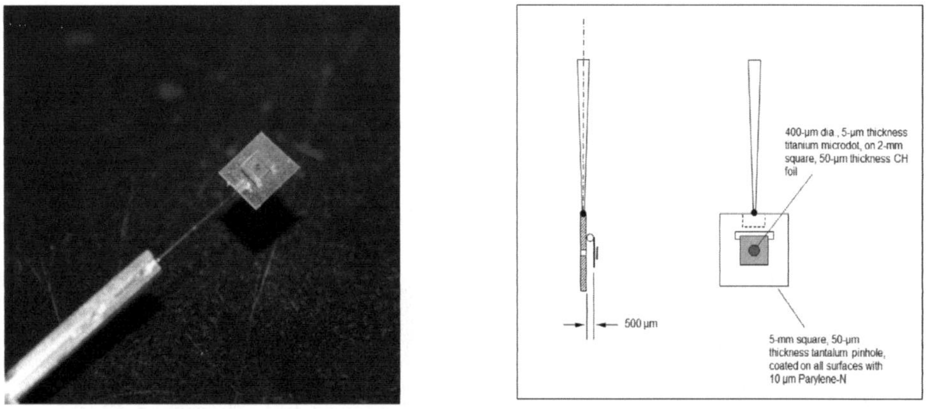

Figure 1. Conventional backlighter target consisting of a titanium micro-dot on a CH film and mounted behind a tantalum pinhole.

2.2. Micron-scale backlighter targets.
The micron-scale backlighter target, as described here, is a markedly different design to that seen in §2.1. It consists of a square-sectioned micro-wire with dimensions of a few μm square x a few tens of μm in length, as shown in schematic form in Figure 2.

Figure 2. Schematic form of the micron-scale backlighter target. Width and depth are a few μm, whilst length may be 10s to 100s μm.

This micro-wire is supported on a low-Z, CH membrane with a thickness of 1μm. This membrane is, in turn, mounted on a Y-shaped support formed from the remainder of the silicon wafer substrate after the manufacturing processes which will be detailed later in §4.

Journal of Physics: Conference Series **713** (2016) 012001 doi:10.1088/1742-6596/713/1/012001

2.2.1. Initial micron-scale backlighter target design.

The computer aided design (CAD) drawing for the initial design is shown in Figure 3. The drawing shows a 100mm diameter silicon wafer substrate with the CH membranes having dimensions of 1mm x 1mm which are located on a pitch of 1.25mm x 1.25mm, giving approximately 1900 targets. The zoomed-in section of Figure 3 shows this in more detail. Also seen on the drawing are diagnostic trenches and alignment marks for use during the fabrication process.

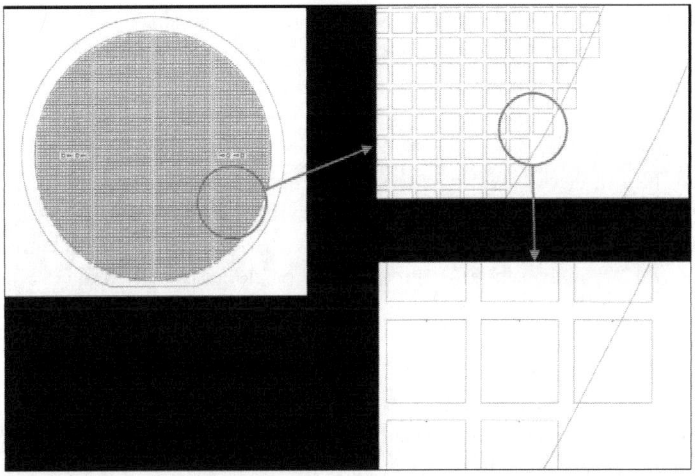

Figure 3. CAD drawing of the 100mm silicon wafer with approximately 1900 targets. The zoomed-in sections show the individual targets in more detail.

2.2.2. Design for first production run.

Prior to the first production run, it was decided to change the design by increasing the CH membrane area to measure 2mm x 1mm to allow easier access to the backlighter for both the laser beam and the alignment optics. The CAD drawing for a single target is shown in Figure 4. Alignment arrows are included in the design to allow the target to be more easily located and positioned when in the target chamber.

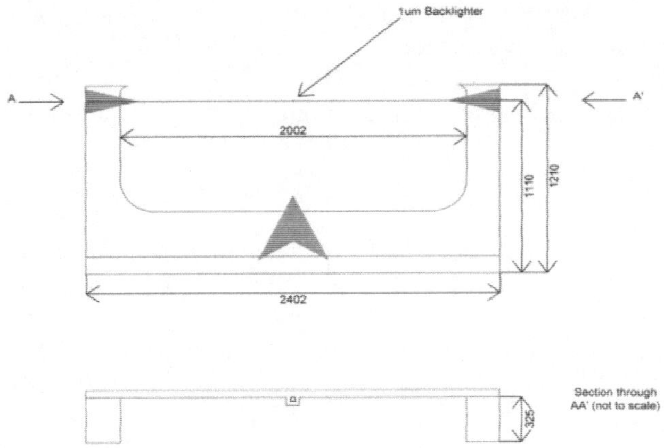

Figure 4. CAD drawing for the first production run of the micron-scale backlighter target. The backlighter is found at the intersection of the vertical and horizontal arrows.

Journal of Physics: Conference Series **713** (2016) 012001 doi:10.1088/1742-6596/713/1/012001

3. MEMS-based fabrication.

These manufacturing technologies offer the opportunity to fabricate targets with significantly smaller dimensions than previously achieved, and with much finer tolerances. Unless requested otherwise, the substrates for this type of target fabrication are usually silicon wafers which are typically 100mm in diameter, have a thickness of 300-500 microns and are flat from edge to edge on the micron scale. MEMS microfabrication techniques are based upon three basic processes: deposition, patterning and etching, and are described briefly below.

3.1. Deposition.

Generally this capability includes sputter-coating, thermal evaporation, chemical vapour deposition, thermal oxide growth and spin/dip coating. Consequently it is possible to deposit precisely controlled films of metals, dielectrics and polymers.

3.2. Etching.

These processes may be either dry etching or wet etching. In the former, a plasma containing highly reactive species (e.g. fluorine, chlorine) is generated which etches the substrate and/or previously deposited thin films. In the latter, the substrate and its coatings are immersed in a solution which is selected to etch the required layer, whilst leaving the other layers unaffected.

3.3. Patterning.

This is the key process and uses either optical or e-beams tools to pattern a thin film of photo- or e-beam sensitive material known as a resist which has previously been applied to the substrate by spin or dip coating. The exposed and developed resist film is then used as a protective stencil through which deposition or etching can take place, thereby transferring the pattern into the substrate or its coatings.

Optical lithography uses photomasks to define the patterns. Masks usually comprise of an extremely flat quartz plate with the pattern defined on one side of that plate in a 100nm layer of chromium. In cases where only low resolution is required a photomask defined on an acetate sheet by high resolution inkjet printing may be used to reduce costs, but are prone to greater wear and tear with repeated use and can suffer from excessive thermal expansion in poorly controlled processing environments.

Electron beam lithography is utilised to define the highest resolution features and uses an e-beam tool which takes the device CAD files (converted into an appropriate format) and uses them to steer the e-beam on the surface of the resist and hence to directly write the required pattern which is then developed prior to use in the next process step.

These basic processes are repeated as often as required, and in a pre-determined order until the final structures are created.

4. Fabrication.

The basic process steps are shown in schematic form in Figure 5. This figure shows the processing in cross-section and focuses on a single target. The patterning processes are able to align the various steps with sub-micron accuracy so that it is relatively straightforward to ensure that the micro-wire is positioned at the edge of the CH membrane, as required – and as shown later in Figure 11. Although the processing appears relatively complicated when compared with conventional target manufacture and utilises sophisticated processing tools, it should be remembered that the process is able to provide micro-wire backlighter targets with sub-micron tolerances and at the fabrication rate of ~1900 targets per wafer, which was not previously possible.

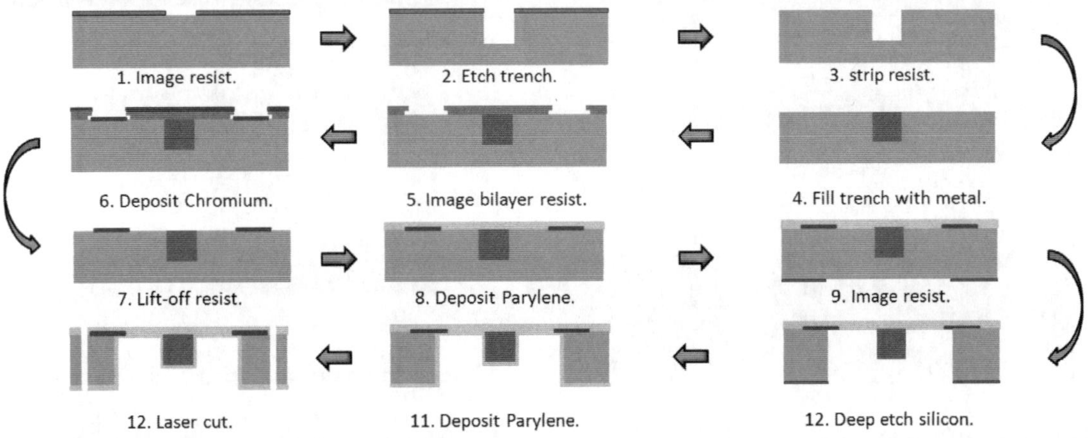

Figure 5. Basic process steps for the manufacture of the micro-scale backlighter targets.

5. Results.

Figures 6 to 11 show the backlighter targets at various stages in the fabrication process. Figure 6 and 7 show the etched trenches (cf. Figure 5 – step 2) whilst Figures 8 and 9 show a wafer after trench filling (cf. Figure 5 – step 4). Figure 9, is an atomic force microscope (AFM) image and demonstrates that the surface of the filled trench is flat to better than 45nm. Figure 10 is following the resist lift-off process (cf. Figure 5 – step 7) and Figure 11 is a finished target (cf. Figure 5 – step 12).

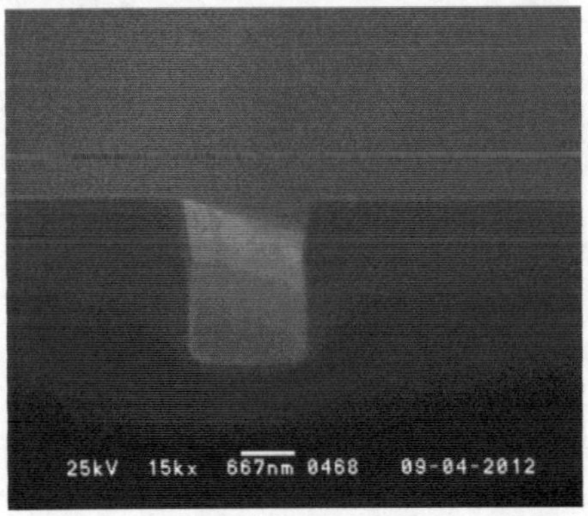

Figure 6. 1μm etched trench in silicon before resist is removed.

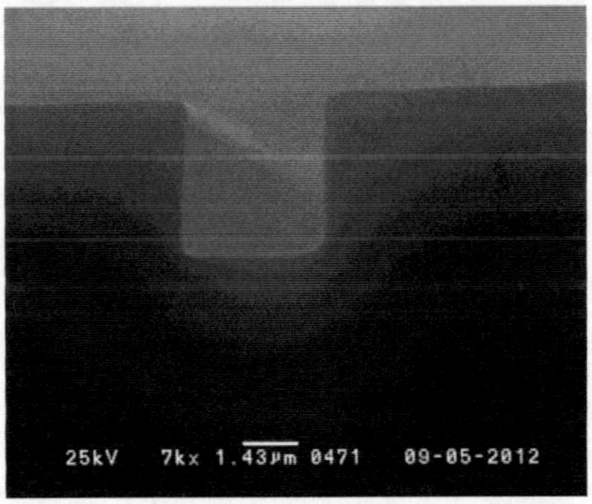

Figure 7. 3μm etched trench in silicon before the resist is removed.

Journal of Physics: Conference Series **713** (2016) 012001 doi:10.1088/1742-6596/713/1/012001

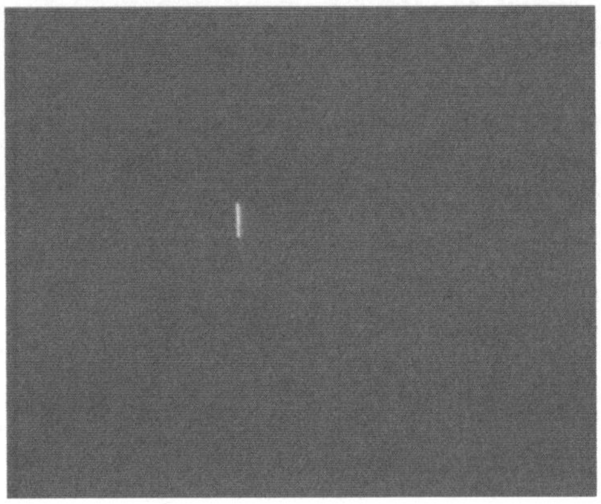

Figure 8. Optical image of filled 1μm x 1μm x 10μm trench.

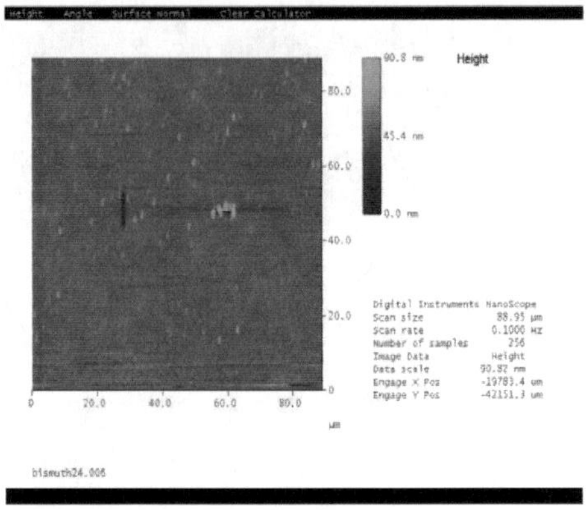

Figure 9. AFM image of a filled 1μm x 1μm x 10μm trench. The filled surface has a flatness of better than 45nm.

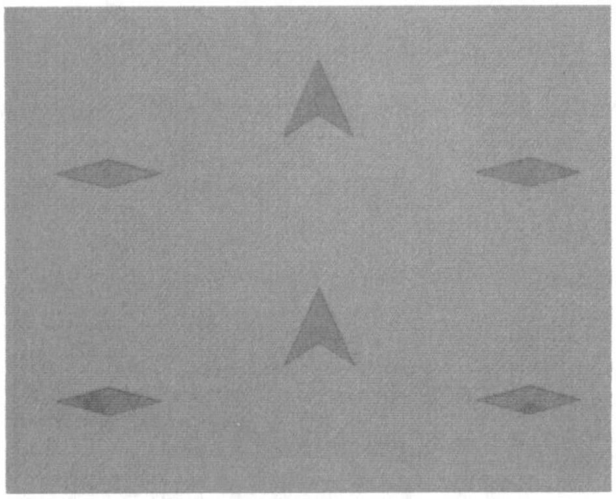

Figure 10. Wider area of the silicon wafer after the lift-off process, showing the Cr alignment arrows.

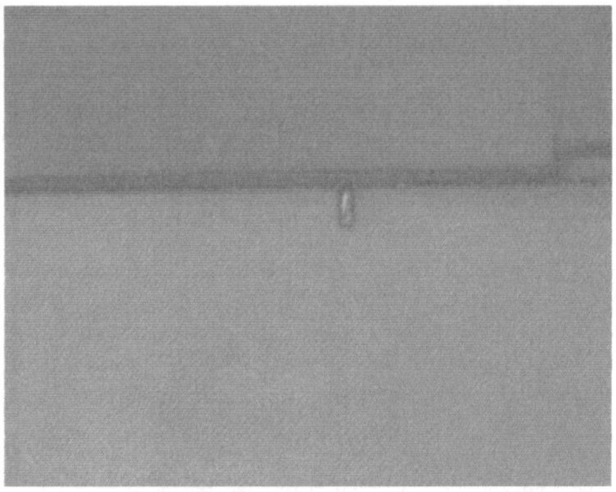

Figure 11. Edge of the CH membrane with a 3μm x 3μm x 10μm target at the edge and ready for use.

6. Conclusions.

In conclusion, we have successfully manufactured micron-scale backlighter targets with three different cross-sectional dimensions: 1μm x1μm, 3μm x 3μm and 5μm x 5μm. They were manufactured using MEMS-based fabrication techniques which are suitable for scaling-up to produce batches of up to a few 10s of wafers, using the currently available equipment. Each wafer potentially provides 1000 – 2000 targets and therefore a batch production of (say) 20 wafers could yield up to 40,000 micro-scale backlighter targets.

High volume fabrication of laser targets using MEMS techniques

C Spindloe[1,2], G Arthur[2], F Hall[2], S Tomlinson[1], R Potter[3], S Kar[4], J Green[1], A Higginbotham[5] N. Booth[1] and M K Tolley[1,2]

[1]Science and Technology Facilities Council, Rutherford Appleton Laboratory, Harwell Oxford, Chilton, Oxon, OX11 0QX, UK
[2]Scitech Precision Ltd, Rutherford Appleton Laboratory, Harwell Oxford, Chilton, Oxon, OX11 0QX, UK
[3]Centre for Materials and Structures, School of Engineering, Brownlow Hill, Liverpool, L69 3GH, UK
[4]Centre for Plasma Physics, Queen's University Belfast, BT7 1NN, UK
[5]Department of Physics, Clarendon Laboratory, University of Oxford, Parks Road, Oxford OX1 3PU, UK

E-mail:christopher.spindloe@stfc.ac.uk

Abstract. The latest techniques for the fabrication of high power laser targets, using processes developed for the manufacture of Micro-Electro-Mechanical System (MEMS) devices are discussed. These laser targets are designed to meet the needs of the increased shot numbers that are available in the latest design of laser facilities. Traditionally laser targets have been fabricated using conventional machining or coarse etching processes and have been produced in quantities of 10s to low 100s. Such targets can be used for high complexity experiments such as Inertial Fusion Energy (IFE) studies and can have many complex components that need assembling and characterisation with high precision. Using the techniques that are common to MEMS devices and integrating these with an existing target fabrication capability we are able to manufacture and deliver targets to these systems. It also enables us to manufacture novel targets that have not been possible using other techniques. In addition, developments in the positioning systems that are required to deliver these targets to the laser focus are also required and a system to deliver the target to a focus of an F2 beam at 0.1Hz is discussed.

1. Introduction

There have been target fabrication activities in the UK for a number of years both at the HELEN laser in AWE and the Vulcan Laser in the Central Laser Facility at Rutherford Appleton Laboratory [1]. These facilities have supported experiments on laser systems that require targets at relatively low volumes. The technologies that have been used to deliver targets to these systems has not vastly changed over time with precision machining, thin film coating, and precise hand assembly being the core activities of a target fabrication group. However since the development of the Vulcan Petawatt laser in 2002 [2], there has been a marked increase in the number of high intensity laser facilities across the world and therefore an increase in the number of shots available. This allows the experimental community to carry out a wider variety of experiments and therefore utilise a wider

Journal of Physics: Conference Series **713** (2016) 012002 doi:10.1088/1742-6596/713/1/012002

variety of target types. Recent laser developments that involve the laser moving from flash lamp to diode pumped designs have led to the possibility of switching from experimental based laser facilities to application based facilities. With this transition from large single shot facilities, such as Vulcan, to much higher repetition rate systems, such as the Astra Gemini laser which is a 2 beam Petawatt (PW) system with 0.5PW in each beam with the ability to operate at rates of approximately 1Hz [3], these systems are no longer shot limited. These application based facilities require large volumes of relatively simple, high specification targets but conventional techniques are not equipped to deal with this demand.

As shot numbers have increased by two orders of magnitude it is clear that to be able to deliver a sufficient number of targets to the laser facilities the existing fabrication techniques need to be expanded. Initially, for simple target geometries, these numbers could be provided using basic fabrication techniques such as 'float off' foils mounted over an array of holes, or using clamp raster mounts to give large numbers of targets in a small puck [4] . Experiments carried out on these target types gave the first indications of application based experiments that could be carried out in the future [5]. There is however a limit to the target geometries that can be produced using these techniques with 2D foil targets being the only viable type that can be produced, and while useful for simple ion acceleration experiments the more complex and scientifically interesting targets cannot be fabricated in high enough numbers to utilise the shots available. It should also be noted that targets, such as 10 nanometer thick gold foils, are extremely fragile and when floated over a hole of 400-500 microns diameter the yield is significantly compromised compared with thicker materials and low enough to make the experiments target resource limited. In some cases this can be overcome by making the mounting aperture smaller but other experiments require specific larger target geometries. Some high-strength target materials, such as Diamond-Like-Carbon (DLC) allow thin films in the range of 2-5 nanometers to be produced [6] and this has improved results but the yield is still low and to fully utilise the laser facilities high repetition rate capabilities requires an almost prohibitive target fabrication resource. It is also apparent that due to the high shot numbers, the fundamental science experiments using simple target geometries are completed within a few months of the facility opening and therefore to undertake more complex experiments a more complex target is required. The cutting edge of conventional target fabrication techniques cannot in its current state deliver complex targets in high volumes.

In addition there is also the challenge of inserting high numbers of targets into the interaction chamber and thence to the laser focus at a speed comparable with the laser shot rate; this includes problems that are encountered when trying to align targets within short timescales. It is clear that any fabrication process that would produce targets that are precisely and reproducibly located to within a micron is ideally suited to these high rep rate applications.

2. MEMS fabrication for laser targets

MEMS fabrication techniques rely on three basic steps: deposition, patterning and etching. These additive and subtractive steps are repeated as often as required, in the appropriate order until the finished device is manufactured. The basic substrate is usually a silicon wafer; the deposition steps are provided by various sputter and thermal coating tools; the patterning by optical or e-beam lithography, as deemed by the resolution required, and the etching is either wet (e.g. acidic or alkaline solutions) or dry (plasma). These processes enable the fabrication of 2 and 2 ½ dimensional structures on the nanometer scale with high accuracy and repeatability in high numbers. A typical example of a MEMS device manufactured in the semiconductor/MEMS industry would be the accelerometers in car airbag safety systems [7] where micro-mechanical parts are integrated with microelectronics. We are now using these technologies to produce a range of targets in higher volumes than previously produced and to a high specification. It should be noted that to produce these targets significant infrastructure is required and includes not only advanced deposition, patterning and etching tools but also high quality

Journal of Physics: Conference Series **713** (2016) 012002 doi:10.1088/1742-6596/713/1/012002

cleanroom facilities to ensure that there is no particulate contamination. This is essential when working with devices that can be of the order of one micron in size as any debris or dirt can damage the targets and render it unusable

2.1. *Simple thin foil targets*
The simplest target that can be produced is a thin foil suspended over a small hole. Such targets can be divided into two categories: 1) Standard thin foil targets such as those used in radiation pressure acceleration experiments in the hole-boring regime have thicknesses in the micron scale and 2) Ultra-thin foils which are used in the light sail regime which are considerably thinner, in the nanometre scale.

Standard thin films require a foil, coated onto a release layer, to be 'floated off' onto the surface of water. The foil is then picked up using a bespoke mount which includes a hole which defines the target area of freestanding foil. This is a labour intensive process and does not lend itself to high target numbers. MEMS techniques allow the yield of such targets to be greatly increased as the plastic foil can be coated directly onto a wafer substrate which will eventually become the target mount.

In the simplest case chemical vapour deposition (CVD) techniques are used to deposit a parylene polymer thin foil directly onto the surface of a wafer. The reverse side of the wafer is then coated with an optical resist, such as AZ9260, and is then placed in a mask aligner for UV exposure through a photo mask patterned with the desired array of target holes. After development the remaining resist protects specific areas of the wafer whilst leaving the other areas exposed for further processing. Using deep reactive ion etching (DRIE) the exposed areas of the wafer can be etched, stopping at the silicon/parylene interface leaving a suspended parylene foil over an array of holes (Figure 1& 2). The geometry and position of this hole can be closely controlled during the design stage of the mask and this allows for accurate and quick target positioning in the target chamber by referencing these drawings. An addition benefit of coating the parylene onto the surface of a polished silicon wafer is that it produces a foil that has a similar surface roughness as the original surface of the silicon wafer. This is an important requirement for certain classes of experiment (TNSA ion acceleration) that rely on the foil being perfectly flat on the scale of the laser focus as the ion beams generated in the acceleration mechanism propagate normal to the target surface.

Figure 1. The etched holes on the rear (polished) surface of a silicon wafer.

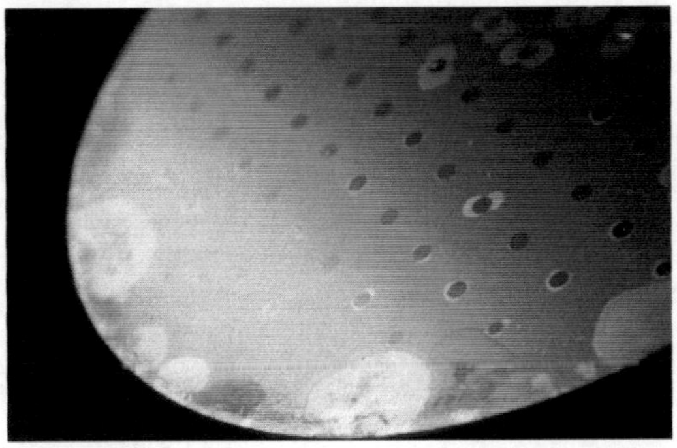

Figure 2. A suspended Parylene (CH) film over open apertures in the silicon wafer.

Characterisation data of at a range of positions on the parylene foil measured using a white light interferometer show the roughness across a random sample is consistent with the surface roughness measured from the unpolished side of the silicon process wafer.

2.2. Ultra-thin foil targets

These targets have applications in ion acceleration experiments where a foil of a few nanometres in thickness is accelerated by a different mechanism, namely 'break-out afterburner' [8] or in induced transparency experiments [9]. Targets are conventionally fabricated using the same techniques as described earlier but due to the thin nature they are inherently more fragile and yield rates can be as low as 5% making them very labour intensive and therefore not an idea target for higher repetition rate experiments and high volume fabrication. MEMS processing techniques can also be used for the manufacture of this type of target, and combined with material choices that exhibit a higher strength than standard foils the yield of the targets can be increased dramatically.

In one such example a thin foil (5nm) of tantalum nitride (TaN) was deposited using an atomic layer deposition (ALD) process onto a silicon wafer [10] with the thickness measured post deposition by ellipsometry to be 4.46nm. ALD was chosen as it is a relatively low temperature technique that provides highly uniform, conformal, pin-hole-free films with excellent control over film thickness. This wafer had an interface layer grown between the bulk and the ALD TaN to allow for etching without the damage of the foil and a resist protection layer was also then placed onto the target material. Subsequent processing produced a large number of targets on a single wafer (figures 3 and 4) with an almost 100% yield.

Figure 3. The silicon wafer with a TaN coating and etched through features.

Figure 4. View showing the 300 micron target aperture on a 2x7mm target mount.

An interesting feature of these target foils is that there is a surface form that is created by etching away the rear of the wafer and exposing the target. This form is dependent on the thickness of the TaN layer with an increase in surface topography observed as the TaN layer becomes thinner. This is illustrated by the data given in table 1, which shows the height of the highest part of the target foil above the surface of the silicon wafer.

Table 1. The difference between the highest part of the target sample and the surface of the silicon wafer for a number of target thicknesses.

Thickness (μm)	Height above wafer (μm)
5nm	14.55
10nm	14.60
15nm	12.40
25nm	2.68

Analysis of the surface form of the targets shows that for a 25nm target there seems to be much flatter and more uniform structure to the target and a smaller deviation from the surface of the silicon wafer. The surface topography is believed to be due to internal stress in the target and further work is required to investigate this.

Journal of Physics: Conference Series **713** (2016) 012002 doi:10.1088/1742-6596/713/1/012002

2.3. *Microdot targets*

In some experiments using facilities such as Free Electron Lasers (FELs) it is important that the laser intensity is uniform across the area of interest, in other experiments the target material of interest may be required to only be very small to ensure that it can be heated efficiently by the laser pulse. Therefore the thin foil is reduced from a continuous area to a micron scale dot. Certain geometries lend themselves to be coated using standard thin film techniques though a mask that can be prepared by, for example, photo etching. There is however a limit to the minimum size of mask that can be produced that can give well defined features, and for thicker coatings a gaussian distribution is observed in the 'dot'. Certain materials are not amenable to being evaporated using directional means such as thermal or electron beam coating and are better sputtered, however this process is not an ideal process for coating through masks and therefore a more refined approach is needed.

An array of 40 x 40 targets, shown in figure 5, giving up to 1600 shots was prepared using these methods. The target design was a 1 micron thick plastic support with a 5 micron diameter, 1 micron thick iron dot suspended in its center. Using a bi-layer resist combined with a number of coating processes and a lift-off process, a well-defined dot can be deposited using plasma sputtering. The bi-layer resist enables the coating to be removed from around the dot without damaging the dot coating and etching by DRIE can then expose the supporting film leaving a suspended dot target shown in figures 6 and 7.

Figure 5. The 40 x 40 array of apertures on a silicon chip

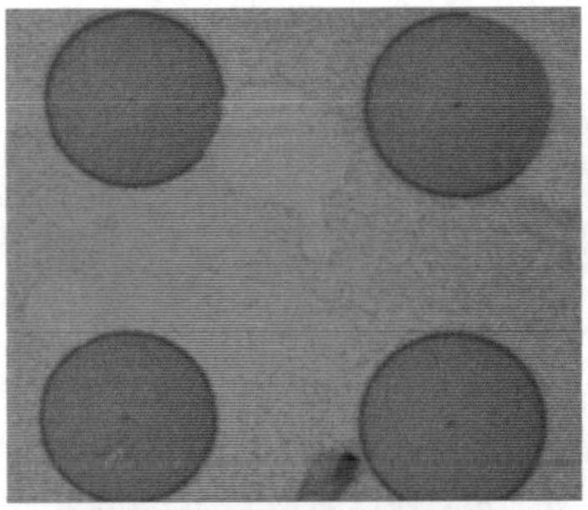

Figure 6. The suspended plastic membrane with the iron dot just visible

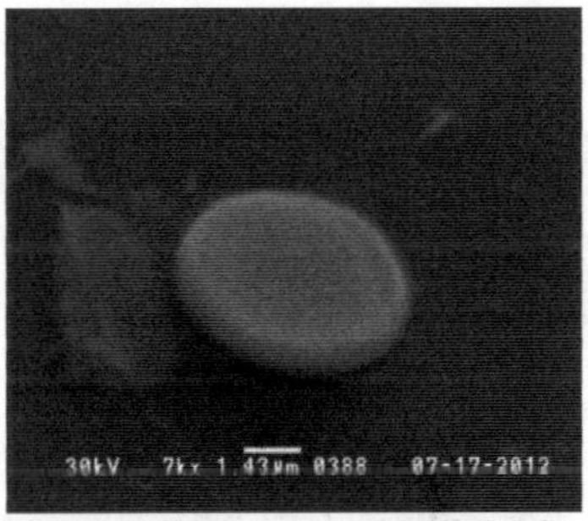

Figure 7. Characterisation of the iron dot showing a good uniformity and size

2.4. *Grating targets*

Gratings can be used in several ways, firstly as targets (e.g. in ion acceleration experiments), but also as diagnostic tools within the target chamber. These gratings have periods which from a several microns down to a few hundred nanometres but by using a range of MEMS processes, including optical and e-beam lithography, deposition, lift-off and deep reactive ion etching, they can be successfully fabricated.

The example shown in figure 8 is a grating structure used in ion acceleration experiments and consists of a 1 micron deep grating in the surface of a 10 micron thick parylene film. The grating consisted of 2 micron lines and 3 micron spaces. The experiment required the sidewalls of the grating to be sloping rather than vertical and this was achieved during processing by heat treatment of the mould from which the final grating was cast.

Figure 8. Cross-section of the mould from which the grating targets were cast. Heat treatment enabled the sloping wall angle to be reliably reproduced.

3. Target Positioning

A technical challenge that is inherently linked to the fabrication of targets in numbers that are suited to the next generation of high repetition rate laser systems is the ability to be able to deliver targets to the laser focus with the required accuracy and speed. There are a number of factors that will define the specification of the target positioning system but the main driver is the f-number of the focussing optic. With a larger f-number the Raleigh range of the system is longer and there is a larger tolerance on the position of the target. It is also important to take into account the type of experiment that is being carried out, for example, in TNSA ion acceleration experiments where the ion species are ejected normal to the target surface, the ability to manoeuvre the target to be perpendicular to the laser beam and therefore the diagnostics is essential. The Central Laser Facility is developing a HAMS positioning system (High Accuracy Microtarget System) for use on the Astra Gemini laser system. This system employs a F2 parabola to deliver a spot size of 2 microns and intensities of 0.5PW in its 2 beams. For each of the factors affecting the positioning of a target within the chamber the design specification has tolerances and accuracies to ensure that a fresh target can be positioned to the laser focus at the required repetition rate of 0.1Hz

3.1. Stages

An X, Z linear translation stage with a tripod attached to the top was used to allow the stages to reach the required positions. The rotation of the target is controlled by a rotation mount that sits on top of the stages. The positioning of a target was agreed to be within an accuracy of +/- 4 microns in the direction of the laser (Z), to ensure it is within the laser focus to obtain the maximum intensity and a positional accuracy of +/- 10um in X and Y was agreed to be enough for most targets. The U and V rotation stages are required to ensure the front plane of the target is perpendicular to the laser and the W rotation is required to rotate the wheel and deliver a new target to the focal position.

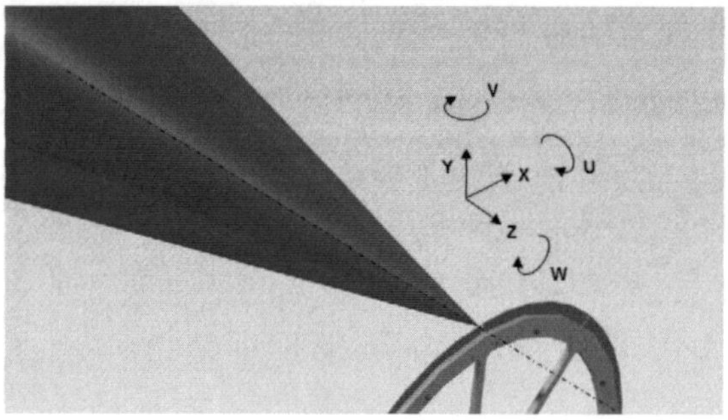

Figure 9. The coordinate system used by the HAMS.

3.2 Interface Wheel

To ensure that the targets can be attached to the stages an interface wheel is needed. This is essentially the target mount. However it can add no sources, or limited sources, of target offset due to its manufacture or the mounting of the targets onto it. Any movement in the Z direction would take the target out of the focal plane and reduce the intensity of the laser incident on the target. A wheel was machined in a ceramic that has good thermal stability and characterised to have flatness over the mounting positions of the target to better than 2 microns. This is within the tolerance that is acceptable for the wheel as long as other sources of error do not accumulate.

3.3 Target Sections

As described earlier a number of geometries can be fabricated from silicon wafers and for the HAMS system a target section is produced that mounts on the interface wheel. This target section can be designed for a number of different target sizes but essentially the targets are patterned around the circumference of the section and then the section is mounted onto the interface wheel. Detailed characterisation of one section shows that it is flat to less than 0.5 microns and 2 sections mounted opposite each other on the interface wheel are flat to within 2 microns.

3.4 Target Alignment

Target alignment is typically achieved at the moment using retro reflections [11] however with higher repetition rates this alignment procedure is not reliable enough to reach the required shot rates due to the fact that for thinner targets imaging the rear surface is very difficult as it becomes more transparent. The CLF is developing a multi wavelength interferometer system [12] using a focussed beam onto the target to allow the positioning of targets with a higher accuracy in Z and is investigating the use of the system to be able to measure any tip and tilt in the target. Further work is being undertaken to implement a feedback loop will that will eventually be able to automate the alignment of the targets and to reach repetition rates of up to 10Hz.

4. Conclusions

As the trend towards applications based systems is increasing, shot numbers have ceased to be the limiting factor in experimental campaigns and target delivery has become the critical path towards the exploitation of the laser technology. We have shown that there are a number of methods that are available using MEMS technology to deliver high aspect ratio target solutions in an effort to deliver these high target numbers however we have highlighted that there is an integrated challenge that does not only include the target fabrication. A fully engineered target stream that is designed to deliver a range of target types within a single delivery suite will allow facilities to deliver experimental campaigns to the user communities or to exploit the shot rates for application based experiments.

Over the next few years we will develop further target geometries, and fully characterise the system both for mechanical performance against specification but also for performance of each part of the HAMS system when under shot conditions.

References

[1] Sutton D W 1987 *J. Vac. Sci. Technol.* A **5,** 2773-77
[2] Danson C N et al 2004 *Nucl. Fusion* **44** 239-246
[3] Hooker C et al 2006 J. *Phys. IV France* **133** 673-677
[4] Spindloe C, Dival E, Schaumann G, Nazarov W and Tolley M K 2009 *Central Laser Facility Annual Report* 301-305
[5] Borghesi M et al 2011 *Proc. of SPIE* **8079** 80791E-4
[6] Ma W, Liechtenstein V Kh, Szerypo J, Jung D, Hilz P, Hegelich B M, Maier H J, Schreiber J and Habs D 2011 *Nuclear Instruments and Methods in Physics Research Section A: Accelerators, Spectrometers, Detectors and Associated Equipment* **655** 53-56
[7] Judy J W 2001 *Smart Mater. Struct.* **10** 1115-34

Journal of Physics: Conference Series **713** (2016) 012002 doi:10.1088/1742-6596/713/1/012002

[8] Yin L, Albright B J, Hegelich B M, Bowers K J Flippo K A, Kwan T J T and Fernandez J C 2006 *Physics Of Plasmas* **14** 056706

[9] Henig A et al 2009 *Phys. Rev. Lett.* **103** 045002

[10] Fang Z, Aspinall H C, Odedra R, Potter R J, 2011, J. Cryst. Grow. 331, (1), p33-39

[11] Carroll D et al 2011 An assessment of the reproducibility of the Gemini retro focusing system *Central Laser Facility Annual Report*

[12] Proc. SPIE 8850, Target Diagnostics Physics and Engineering for Inertial Confinement Fusion II, 885002 (26 September 2013);

Production of a thin diamond target by laser for HESR at FAIR

F Balestra[1,2], S Ferrero[1], R Introzzi[1,2], F Pirri[1], L Scaltrito[1] and H Younis[3]

[1]Department of Applied Science and Technology, Politecnico di Torino, c.so Duca degli Abruzzi 24, 10129, Torino, Italy
[2]INFN- Sezione di Torino, c/o Politecnico di Torino, c.so Duca degli Abruzzi 24, 10129, Torino, Italy
[3]Department of Physics, COMSATS Institute of Information Technology, Park Road, 44000, Islamabad, Pakistan

E-mail: francesca.balestra@polito.it

Abstract. In the future hadron facility FAIR, the HESR ring will supply antiprotons in the momentum range 1.5-15 GeV/c as projectiles to study charm, strangeness and a wide range of other Physics topics. For all these reactions it will be necessary to use internal targets and in particular, for the production of systems with double strangeness, a solid ^{12}C target will be used. Inserting a solid target inside an antiproton ring creates two main problems: a large background on the detectors due to the overwhelming amount of annihilations and a strong depletion of the beam due to all the hadronic and Coulomb interactions of the antiprotons with the ^{12}C nuclei. The width of the target plays a crucial role in minimizing these unwanted effects. Two wire-shaped prototypes have been already realized, starting from a thin diamond disk. The wire shape has been obtained by using a femto-edge laser. One prototype has been submitted to irradiation by protons of 1.5 MeV and to simultaneous Back-Scattering control to test the impurity level, the ^{12}C density, the radiation hardness and possible phase modifications during irradiation. Both the prototypes have been submitted to Micro-Raman spectroscopy in order to scan the carbon phases along the width. The results show performances which satisfy the experimental requirements.

1. Introduction – The $\bar{\mathrm{P}}$ANDA experiment

The $\bar{\mathrm{P}}$ANDA experiment (anti-Proton Annihilation at Darmstadt) is one of the most important projects at the facility FAIR under construction in Darmstadt, Germany. $\bar{\mathrm{P}}$ANDA will exploit the SIS100 synchrotron of FAIR, able to provide intense pulsed ion beams (from p to ^{92+}U), to create antiprotons, starting from a primary proton beam. The antiproton ($\bar{\mathrm{p}}$) beam will be injected into the High Energy Storage Ring (HESR) and will collide at each round with the fixed target inside the $\bar{\mathrm{P}}$ANDA setup.
The scientific program of this experiment covers several Physics topics that range from the gluonic excitations to the physics of strange and charm quarks and nucleon structure [1]. In particular, the Physics of the strange and doubly strange systems is very attractive since it is largely unexplored.
The doubly strange systems, to be studied, are:

a) Ξ^- hyperatoms, which can give information on the Ξ^--nucleus potential [2],
b) Ξ^- hypernuclei, which can provide information about the Ξ^--nucleon interaction,
c) $\Lambda\Lambda$ hypernuclei, which can supply information on the hyperon-hyperon interaction [3].

The production of these systems, in $\bar{\text{P}}$ANDA, takes place through a two-step process:

a) interaction of an antiproton with a nucleon to produce a Ξ^- hyperon,
b) capture of the hyperon by another atom, after slowing down and stopping, with the formation of the doubly strange system [4].

Two targets are necessary for this purpose: in a primary target, inside the beam pipe, the Ξ^- hyperon is produced by an antiproton; in a secondary target, outside the beam pipe, the hyperon goes to stop and is captured.

In the following, the primary target features, the technique to produce it and the tests performed on the realized prototypes will be described.

2. The primary target for $\bar{\text{P}}$ANDA: features and realization

The production of Ξ^- by antiproton annihilation occurs through the reaction $\overline{p}+N \rightarrow \Xi^- + \overline{\Xi}$, whose cross section is very low ($\approx$ 2 μb). By locating an "internal target" inside the ring, it is possible to reuse, in the following round, the antiprotons that did not interact in the previous passage through the target. In this way the bunch is more effectively used with respect to an extracted beam (see figure 1).

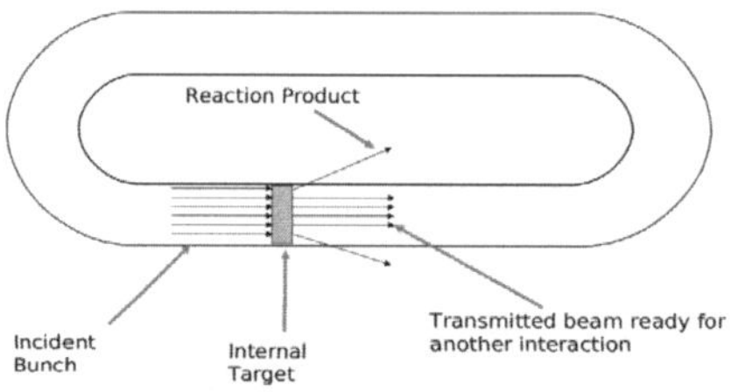

Figure 1. Scheme of a target inside a ring: not interacting projectiles can be reused.

A suitable choice for the $\bar{\text{P}}$ANDA internal target is the ^{12}C nucleus. With this choice the hyperons are significantly slowed down in the dense nuclear matter of the residual nucleus after their production [4]. Nevertheless, a solid target inside a ring creates several problems to the beam, to the detectors and to the target itself. In fact: a) the bunch will undergo a strong depletion due to the hadronic and Coulomb interactions; b) a too high annihilation rate can blind the detectors; c) the interaction of the antiprotons can generate high thermal and mechanical stresses and possible charge pile-up inside the target. Therefore, it is mandatory to properly choose material, shape and dimensions of target in order to minimize these problems.

A very good mechanical resistance as well as large thermal conductivity is assured by a diamond target. These features are important because they make the target strong enough to resist to beam impact stresses, they allow to easily handle it inside the beam pipe and permit to dissipate the heat produced in the target by the passages of the beam through it.

The shape chosen is a wire of 3 μm thickness and 100 μm width. The small dimensions together with the steering of only the periphery of the beam spot onto the target [5] allow to reduce the annihilation rate down to a tolerable value ($\approx$ $5 \cdot 10^6$ annihilation/s, for the hypernuclear scientific program) for the $\bar{\text{P}}$ANDA detectors [6]. Finally, the Si ring holding

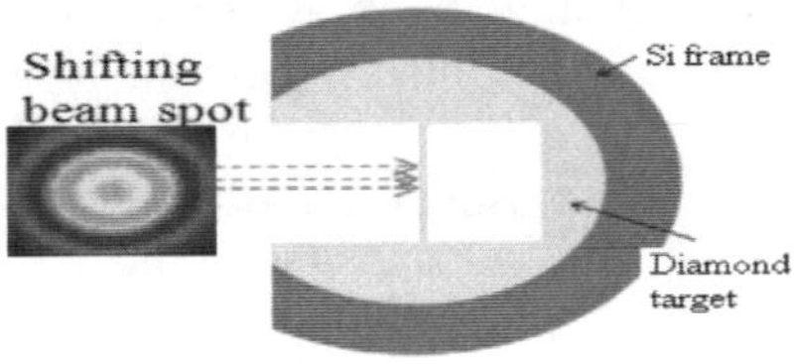

Figure 2. Scheme of the C-shaped support of the target.

the wire target will be cut C-shaped (see figure 2) in order to allow the beam to be slowly steered from left to right avoiding annihilations in the thick frame.

The first attempt to produce the diamond wire was performed by using an amorphous deposition. A carbon film nearly 3 μm thick was deposited on a disk shaped Cu substrate (500 μm thick, 12 mm in diameter). Purity and density were measured by means of the Back Scattering technique (BS), using a 1.5 MeV proton beam at LNL (National INFN Laboratories of Legnaro) at a scattering angle of 165°. The measured areal density of the film was $8.3 \cdot 10^{18}$ atom/cm^2 and the ^{12}C purity level was 54%. A high percentage of impurities was present: 7% of ^{16}O and 39% of H. This impurity level was considered too big because the hydrogen nuclei do not slow down the hyperons and the efficiency in the double hypernuclei formation would be strongly decreased. Moreover, the density of the film was four times less than that of graphite. This means that the amorphous deposition should be four times thicker in order to obtain an equivalent target thickness; however, in such a kind of films higher thickness produces higher brittleness that cannot be tolerated. Finally, the five trials made, using a 30 KeV Ga-ion beam, to cut away two rectangles inside the film, in order to shape a wire, were unsuccessful with final breakage of the wire. Therefore, this technique was discarded [7].

Figure 3. The diamond disk (transparent area) stuck onto the silicon ring (black ring).

A second method was explored. The prototype production started with the manufacturing, from DIAMOND MATERIALS GmbH Company (Freiburg), of disk-shaped diamond films by Chemical Vapor Deposition (CVD) on a Silicon plate of 15 mm diameter and (500 ± 100) μm thickness. An internal disk of lower diameter (11 mm) was then etched away from the Si support obtaining a freestanding diamond plate on the silicon ring, with 15 mm outside diameter and diamond disk thickness of (3 ± 0.5) μm. One of the prototypes is shown in figure 3. The presence of a stressed region is visible as a wavy surface close to the internal border of the silicon ring. This stress could be a source of brittleness and can cause problems in the cutting operations.
This sample was submitted to several tests: the first one was the "micro beam" test at LNL. The diamond disk was exposed to irradiation by a 1.5 MeV proton beam in two points, close to the center of the disk, for about one and two hours, respectively. The beam spot areas were 3470×3470 and 870×870 μm^2. The proton back scattering rate was measured, during the exposure time, in the same conditions (energy and angle) of the test on the amorphous

film. The irradiation results show that the percentage of ^{12}C in the sample is about 99.9% (as expected from the company specifications) and the content of other nuclei is compatible with zero. The areal density is about $50 \cdot 10^{18}$ atom/cm^2, nearly 50% higher than the graphite density (approximately $34 \cdot 10^{18}$ atom/cm^2). Moreover, the density has been found to be equal in both points, irradiated with different beam intensities. This indicates that such an irradiation does not produce significant changes in the diamond structure.
The target prototype was also tested under a multi-cycle low energy proton beam at LNL, to test the response to mechanical stresses (due to the target-beam interaction) similar to the in-ring conditions. The proton beam current was chosen to reproduce the worst energy loss conditions in $\bar{\text{P}}$ANDA ($dE/dx \approx 1.8$ MeV/(g/cm^2) for antiprotons at 3 GeV/c at HESR, while $dE/dx \approx 111$ MeV/(g/cm^2) for protons of 2.75 MeV/c at LNL). The proton current of about 2.7 nA is producing a total energy loss per time unit equivalent to the $\bar{\text{P}}$ANDA antiproton beam. The irradiation was performed at even higher current (4 nA), in order to test the sample in worse conditions than at HESR. As before, the sample was simultaneously monitored measuring BS. The plot of the number of back scattered protons versus time showed a linear trend, indicating that neither structural variation nor charge pile-up occurred.

The second step in the target production was the diamond cut into the wire shape. This procedure was obtained through laser cuts. A first wire about 100 μm wide, 3 μm thick and 6 mm long was successfully obtained by cutting away two rectangles in the disk (see figure 4). A "Femto Edge" type laser of 3 W power, 1064 nm wavelength and 100 fs pulse duration was used for the cut. The effects of the laser cutting on the internal structure of the wire were

Figure 4. The diamond disk after laser cutting: the wire is visible in the centre between the two white rectangles.

investigated through Raman spectroscopy. Three points have been analyzed along the width of the wire: the center, the periphery and the border. A schematic detail of these regions is reported in figure 5 (left) together with the results of the Raman spectroscopy (right). In the figure it can be seen that the bottom curve (center) shows only a clear peak in the region of the diamond (D) phase. The middle curve (periphery), instead, shows a very slight enhancement in the region of the graphite (G) phase (with corresponding reduction in the D phase) and finally the top curve (border) shows a dominance of the G phase with respect to the D phase. The laser cut, therefore, seems to affect the lateral boundary (along the cut edge) of the wire, creating a graphitization phase and leaving the central region unaffected. The graphitization process increases the electrical conductivity at the border. This unexpected phenomenon is actually positive, since the passage of antiprotons through the target could produce electrostatic charges and a conductive channel would create a discharging path toward the supporting frame of the ring. It is important to remark that, in the central part of the wire, the diamond structure does not change and the thermal and mechanical properties remain intact.
A second prototype of diamond wire was produced (by χ*Lab*, Chivasso, Italy) with the aim of verifying the change of structure diamond-graphite, after the cut, on a larger width (≈ 145 μm). The laser system PHAROS (LigthCOnv), a femtosecond, Ytterbium doped, solid state laser with 1.5 W power, 343 nm wavelength and 220 fs pulse duration, was used for the process.

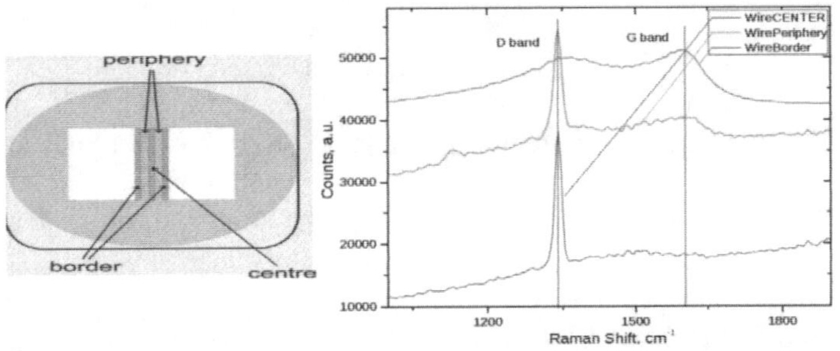

Figure 5. (Left) schematic close up view of the wire profile, showing the three regions examined by Raman spectroscopy. The distances from the edge are, for the border, 2–3 μm and for the periphery and centre about 10 μm and 50 μm, respectively. (Right) Raman spectra of the central (lower curve), peripheral (middle curve) and boundary (upper curve) points along the width of the wire. The D band indicates the diamond while the G band indicates the graphite phase.

The Raman spectroscopy was performed on this wire in three different points, shown in figure 6 (left). The rightmost part of figure 6, instead, reports the Raman spectra measured for the

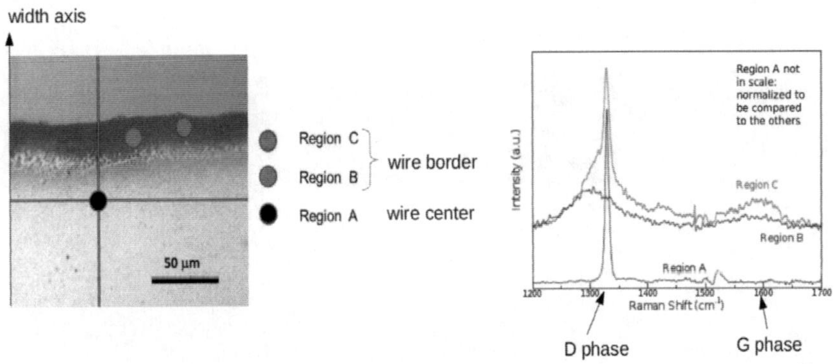

Figure 6. (Left) optical image of investigated points. (Right) Raman spectra measured on the wire center (black curve) and border (blue and red curves).

three different points: A (wire centre), B and C (wire border, close to the cut region). Region A shows the spectrum of pure diamond: it can be seen that no other features appear but a single narrow peak at 1332 cm^{-1} (2 cm^{-1} FWHM). In region C, instead, three features appear in the Raman spectrum: the diamond peak at 1330 cm^{-1} (10 cm^{-1} FWHM), the G band at 1585 cm^{-1} (83 cm^{-1} FWHM), and the D band at 1321 cm^{-1} (93 cm^{-1} FWHM). These results suggest the presence of residual clusters of diamond within the graphitized column, which shows the typical features of nano-crystalline graphite (nc-G). Data obtained in region B, finally, show a complete removal of diamond content and the typical nc-G spectrum. A new laser treatment with different parameters could improve the graphitization process along the wire border.

3. Conclusions and outlooks

A technique to produce a wire shaped primary target, to be inserted inside the $\bar{\text{p}}$ HESR ring, satisfying both the beam constraints and the requirements of the $\overline{\text{P}}$ANDA experiment, has been realized. The diamond wire is fabricated in a two-step process: starting from a diamond disk, obtained through CVD on a Silicon ring, the diamond wire is then produced from the disk

Journal of Physics: Conference Series **713** (2016) 012003 doi:10.1088/1742-6596/713/1/012003

through laser cutting.
Two solid, thin, wire-shaped ^{12}C targets have been already realized.
The first prototype has been submitted to some preliminary tests at the LNL: proton irradiation with BS technique and Raman spectroscopy. These tests have given preliminary positive results: on the disk, mechanical resistance tests and BS controls pointed out that no structural variation and no charge pile-up were created by the irradiation processes; on the wire, Raman spectroscopy highlighted that the laser cutting produced a graphitization process along the sides of the wire. On the second prototype, instead, only a Raman spectroscopy measurement has been performed until now: in this case also, Raman spectroscopy underlined the presence of a spectrum indicating a nc-G structure in the wire border as a consequence of the laser cutting.
Systematic tests of the same type are planned for both the wire prototypes in order to scan the whole target surface; heavier exposures to radiations are planned too. Other prototype realization is foreseen.

References

[1] W. Erni et al. 2013 *Eur. Phys. J.* A **49** 25
[2] F. Iazzi et al. 2009 *Hyperf Interact* **193** 89
[3] F. Iazzi 2014 *EPJ Web of Conferences* **71** 00056
[4] F. Ferro et al. 2007 *Nucl. Phys.* A **789** 209
[5] R. Introzzi et al. *these proceedings*
[6] H. Younis et al. 2013 *J. Radioanal. Nucl. Chem.* **299, issue 2** 951
[7] H. Younis et al. 2014 JINST **9** P04012

Journal of Physics: Conference Series **713** (2016) 012004 doi:10.1088/1742-6596/713/1/012004

The antiproton interaction with an internal ^{12}C target inside the HESR ring at FAIR

R Introzzi[1], F Balestra[1], A Lavagno[1], F Scozzi[1] and H Younis[2]

[1] Department of Applied Science and Technology, Politecnico di Torino, c.so Duca degli Abruzzi 24,10129, Torino, Italy and INFN- Sezione di Torino, c/o Politecnico di Torino, c.so Duca degli Abruzzi 24, 10129,Torino, Italy

[2] Department of Physics, COMSATS Institute of Information Technology, Park Road, 44000 Islamabad, Pakistan

E-mail: riccardo.introzzi@polito.it

Abstract.

In order to fulfill the goal of producing higher rates of doubly strange hyperons, the $\bar{\text{P}}$ANDA collaboration will use the antiproton ring HESR at the future facility FAIR. The low energy hyperon production by an antiproton beam requires to insert a solid target inside the ring. Unwanted side effects of such an insertion are the overwhelming amount of annihilations, which would make the detectors blind, and the fast depletion of the bunch, which circulates inside the ring. The choice of the target material impacts the hyperon production yield: Carbon turned out to provide enough initial hyperon deceleration and keep secondary interactions below a tolerable level. The use of a very thin Diamond target, together with beam steering techniques, seems to be a satisfactory solution to the above problems and will be described hereafter.

1. Introduction

The understanding of the Strange Systems is attracting more and more the interest of the scientific community in the last years. This is due to two main aspects: on one hand the doubly strange systems represent a feasible way to investigate the interactions of strange matter with nucleons and nuclei, on the other hand the technological advancements on intense beams at new hadronic facilities allow a rich production of doubly strange hyperons as at J-PARC (Japan Proton Accelerator Research Complex) and FAIR (Facility for Antiproton and Ion Research).

Exotic Ξ^- atoms, Ξ^- hypernuclei and $\Lambda\Lambda$ hypernuclei (also called Double Hyper-nuclei, DH) are the three possible types of strange systems, in which hyperon-nucleus, hyperon-nucleon and hyperon-hyperon interactions take place.

The investigation of these doubly strange systems allows to shed light on the interaction of a doubly strange hyperon with the nuclear potential in the peripheral regions of the nucleus and with single nucleons [1]. Moreover, the spectroscopy of a double hypernucleus can provide information about the hyperon-hyperon potential and interactions of s quarks[2].

In spite of their importance the available data about these systems are very scarce: only six Ξ^- hypernuclei, seven $\Lambda\Lambda$ hypernuclei and no Ξ^- hyperatoms have been observed up to now.

This lack is due to the low probability of producing Ξ^- hyperons (e.g. in antiproton-proton collisions the cross section is $\approx 2\,\mu$b) and to the short lifetime of the hyperon ($\tau = 1.64 \cdot 10^{-10}\,$s).

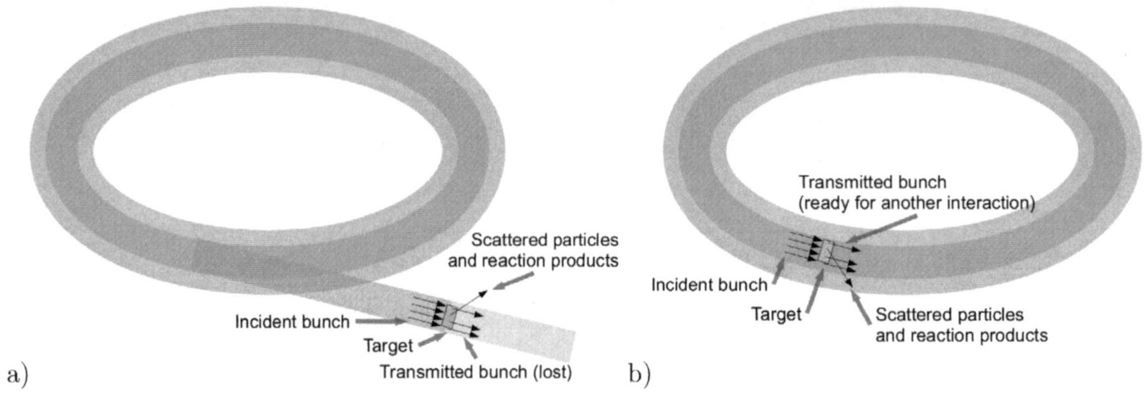

Figure 1. a) extracted beam and b) in-ring target configurations are shown for comparison; in the first case non-interacting particles are lost after crossing the target section while they are reused in the second one.

At the future facility FAIR the ring HESR (High Energy Storage Ring) will supply intense antiproton beams in the momentum range 1.5 – 15 GeV/c [3], which will be exploited by the experiment P̄ANDA (antiProton ANnihilation at DArmstadt) to collect rich statistics thanks to new production and detection techniques.

The reaction antiproton-carbon has been found to satisfy all the requirements to obtain a high rate of double hypernuclei [4]. For this goal the Collaboration P̄ANDA performed a feasibility study of a new experimental technique based on the use of two different targets for the creation of Ξ^- hyperons and of double hypernuclei. One of these targets should be inserted inside the HESR in order to re-use at each round the antiprotons, which did not interact in the previous one. In this way the effect of the low Ξ^- production probability and short lifetime is partially recovered (see Fig. 1). The idea of inserting a solid target inside a storage ring (in particular a ring of anti-particles) is innovative and presents several problems [5].

2. P̄ANDA experiment

Hydrogen jet or pellet targets are commonly used in High Energy Physics. Their interactions with the circulating beam affect the time and spatial structure of the projectile bunch and must be taken into account in managing the beam. In case of the P̄ANDA hypernuclear program, the internal target is required to be solid and denser than gaseous and liquid hydrogen. A nuclear target different from hydrogen is crucial to provide the initial strong Ξ^- deceleration by re-scattering inside the dense nuclear matter. Of course, this greatly increases the unwanted side effects on the antiproton beam. Drawbacks are of two types.

(i) Depletion of the antiproton bunch at each round, due to:
- hadronic interactions with the target nuclei:
  - annihilations,
  - elastic and inelastic nuclear scattering,
  - CEX (Charge EXchange);
- SCS (Single Coulomb Scattering) with the target nuclei;
- Touschek effect in the beam (not depending on the target), negligible with respect to SCS and hadronic interactions;
- Straggling effect;
- Multiple Coulomb Scattering in the target (recovered by the cooling system of the ring if the target is sufficiently thin);

(ii) Target radiation damage, which can produce:
- heating of the target;
- mechanical stress due to the interaction with antiprotons;
- fragmentation of target nuclei with consequent embrittlement;
- accumulation of electrostatic charges, which can perturb the antiproton trajectories.

Type (i) effects require thin and narrow targets to be cared, while type (ii) ones create more problems the smaller the target is. Therefore, the sizes and the features of the target and its beam-spot overlay play a crucial role. To provide feasible measurement conditions, a proper design of the internal target and a careful study of the space-time structure of the beam in the ring are mandatory.

As previously discussed, the interaction of the antiproton beam with the target will generate high mechanical stress, embrittlement, heat and possibly charge pile-up; thus, the target structure and material must feature very good mechanical resistance, as well as large thermal and electrical conductivities.

The reaction of interest, $\overline{p}+N \rightarrow \overline{\Xi}+\Xi^-$, has a maximum cross section close to 3 GeV/c and a threshold at 2.65 GeV/c, while the π production comes into play above 3.1 GeV/c. By using antiprotons at 3 GeV/c, the Ξ^- production is maximized and only two particles are present in the final state.

Another important issue, when using an internal target within a storage ring, is the refill and measurement cycle. The initial multiturn injection of antiprotons, with a momentum of 3.7 GeV/c, is followed by beam cooling and deceleration in order to meet the specification on the antiproton momentum of 3 GeV/c for the optimal Ξ^- production. The measurement phase starts with 10^{10} antiprotons in a rich bunch (about 200 m long), circulating in the racetrack-shaped ring (about 550 m long) at a frequency $f_{ring} \approx 0.5$ MHz.

Direct exposure of the internal target to such an intense beam can create a high annihilation rate, with the risk of blinding all detectors and of wasting many antiprotons. Furthermore, the beam lifetime would shorten due to type (i) effects.

In particular, the initial bunch content would produce higher annihilation rates than the maximum tolerable by the $\bar{\text{P}}$ANDA detectors in the hypernuclear program, $R^{MAX} \approx 5 \cdot 10^6$ ann/s.

A solution to these problems has been found by overlapping only partially the beam spot with the target. In order to reduce the number of interactions at each passage through the target, only the periphery of the bunch (whose radial distribution is nearly Gaussian) will overlap the target. This will be achieved slowly steering the beam toward a wire-shaped, thin, solid target (see Fig. 2). In this way the interaction rate can be controlled to maintain it close to an optimal level.

A proper choice of the material must also account the dependencies of type (i) cross sections on the mass A and atomic Z numbers of the target.

Summing-up, the initial Ξ^- deceleration requires a high enough mass number while keeping unwanted interactions below an acceptable level sets a limit on the same quantity. The best compromise has been found in Carbon [4].

Diamond turns out to be the best material to comply also with mechanical and heat specifications [6]. To the best of our knowledge, the production of alternative Carbon materials, like multilayer graphene samples and nanotube bundles, in a suitable size is still out of reach for nowadays technology. A target has been prototyped with CVD techniques: a free-standing diamond membrane has been produced on a Si ring [6]. The method ensures high purity (99.9%), homogeneous thickness (3 μm) and areal density ($5 \cdot 10^{19}$ cm^{-2}). Then the target prototype has been wire shaped (100 μm wide) by femto-edge laser cut (1064 Å, 3 W).

In each cycle, during the data taking phase, the bunch is depleted at each passage through the target. At the n-th round, the bunch content is given by I_n antiprotons; only a fraction

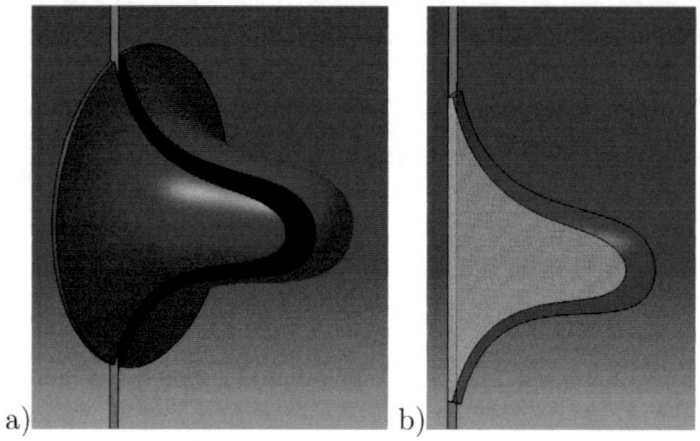

Figure 2. Picture of the steering technique: a) the circular Gaussian represents the beam distribution overlapping the wire target with its peripheral region where the antiproton density is lower; b) the distribution slice, corresponding to the beam-target overlap, is shown.

N_n of them ($N_n = f_n I_n$, $0 < f_n < 1$), will go through the target thin wire, depending on the relative position of the beam spot.

Among the N_n antiprotons a smaller subset, N_n^i, will actually interact with the target through one of type (i) processes. Labeling by the index α each interaction (α = annihilations, elastic nuclear scattering, inelastic nuclear scattering, CEX, SCS, Ξ^- production, ...), the number N_n^α of antiprotons undergoing interaction α is given by:

$$N_n^\alpha = N_n \frac{\sigma_\alpha^A}{\sum_\alpha \sigma_\alpha^A}(1 - e^{-\Sigma}) \approx N_n \frac{\sigma_\alpha^A}{\sigma_{SCS}^A + \sigma_h^A}(1 - e^{-\Sigma}) \tag{1}$$

where σ_α^A are the cross sections of each interaction, $\sum_\alpha \sigma_\alpha^A$ is the total cross section, which is approximated with the sum of the dominant ones by SCS and hadronic processes. The term Σ accounts for all factors coming into play in the exponential decay of the antiproton flux through the solid target thickness, l:

$$\Sigma = \frac{\rho N_{Av}}{A}\left(\sum_\alpha \sigma_\alpha^A\right) l \approx \frac{\rho N_{Av}}{A}(\sigma_{SCS}^A + \sigma_h^A) l \tag{2}$$

where ρ is the target mass density, N_{Av} the Avogadro number, A the target nucleus mass number.

The Single Coulomb Scattering and hadronic cross sections, σ_{SCS}^A and σ_h^A, of the antiproton-nucleus interactions can be obtained from the corresponding elementary antiproton-nucleus ones, σ_{SCS} and σ_h, scaling by Z^2 (SCS) and $A^{2/3}$ (hadronic) respectively. The result is:

$$\sigma_{SCS}^A = Z^2 \sigma_{SCS} \approx 1030\,\mathrm{mb} \qquad \sigma_h^A = A^{2/3}\sigma_h \approx 367\,\mathrm{mb} \tag{3}$$

The Ξ^- production cross section turns out to be lower [9].

$$\sigma_{\Xi^-}^A = A^{2/3}\sigma_{\Xi^-} \approx 10.5\,\mu\mathrm{b} \tag{4}$$

The adoption of the steering technique is meant to mantain N_n almost constant. This condition implies that each rate of type (i) reactions does not vary with time, as well as the depletion rate of the bunch, given by $f_{ring} \sum_\alpha N_n^\alpha$. In particular the annihilation rate remains at the same level as well as the Ξ^- hyperon production rate. The aim of the experiment is a Ξ^- production rate of about 100 per second.

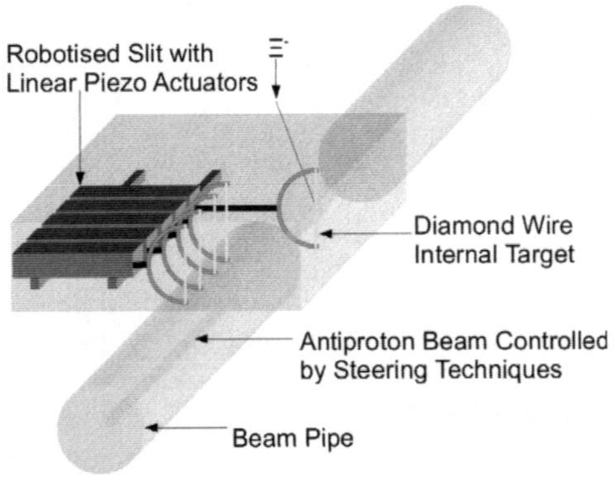

Figure 3. A transparent view of the in-vacuum interaction area shows the inner target array together with its robotised holder. Beam depletion requires a periodic refill of the ring. During antiproton injection the target is moved aside from the beam line by an automatic system based on linear piezoelectric actuators and then set back. An array of diamond wires allows to change the target when degraded by radiation damage.

The diamond atom density of the primary target, $\rho_D \approx 5 \cdot 10^{19}\,\mathrm{cm}^{-2}$ (much higher than the one of liquid hydrogen: $\rho_{LH_2} \approx 4 \cdot 10^{15}\,\mathrm{cm}^{-2}$), helps to reach this goal [10]. The expected ratio of stopped Ξ^-, with respect to the produced ones, turns out to be $R_{\Xi^-} \approx 2.2 \cdot 10^{-3}$ while the stopped Ξ^- fraction that produces $\Lambda\Lambda$ hypernuclei would be $R_{\Lambda\Lambda/\Xi^-} \approx 5 \cdot 10^{-2}$. This will provide about $2.5 \cdot 10^4$ $\Lambda\Lambda$ hypernuclei per month.

A reliable system to precisely move the target to the desired position inside the beam pipe has to be built. The particle injection and measurement cycle within the storage ring requires to set the target along the beam line only during the measurement time and to remove it in the refill time. Furthermore, the target itself will degrade due to the nuclear reactions taking place under the exposure to the beam and it must be substituted from time to time in a quick way. Figure 3 shows the sketch of the target region with an inner view of the beam line where the internal target array and the robotised holder system are positioned [11].

The device must be compatible with high vacuum. Feedback should be provided for external positioning control as well as the possibility to carry different sensors and electric signals intended to monitor the target itself. To satisfy all these constraints, the best solution has been found in linear step motors based on piezoelectric actuators.

3. Conclusions

Hadronic systems with double strangeness are interesting topics of particle Physics. New experiments at FAIR and J-PARC are aiming to investigate them to overcome the present scarceness of data. In particular, the use of antiproton beams at FAIR seems to be a promising technique for doubly strange hyperon production.

This technique requires inserting a thin solid target inside a ring with an intense beam, leading to severe constraints.

The proper design of the dimensions and features of the target and the beam-spot overlap play a crucial role to provide feasible measurement conditions. A wire shaped diamond target, which overlaps only partially the beam spot, together with the beam steering technique will be used.

A robotic control of the target within the beam pipe will be implemented to allow the measurements and the ring-refilling stages.

References

[1] Iazzi F and Szymańska K 2009 *Hyperf. Interact.* **193** 89
[2] Iazzi F 2014 *EPJ Web of Conferences* **71** 00056
[3] Iazzi F et al. 2011 *Proc of Sci.*(STORI11) **020**
[4] Ferro F et al. 2007 *Nucl. Phys. A* **789** 209
[5] Introzzi R et al. 2014 *EPJ Web of Conferences* **73** 05004
[6] Balestra F et al. in the present proceedings
[7] Lehrach A et al. 2006 *Nucl. Inst. Meth. A* **561** 289
[8] Bendiscioli G and Kharzeev D 1994 *Nuovo Cimento* **17** 1
[9] Musgrave B et al. 1965 *Nuovo Cimento* **35** 735
[10] Iazzi F 2008 *Few Body Systems* **43** **1** 97
[11] Sanchez-Lorente A 2013 *J. of Phys.* **426** 012030

Fabrication and characterization of thin polymer targets for laser-driven ion acceleration

A Tebartz, S Bedacht, G Schaumann and M Roth

Institut für Kernphysik, Technische Universität Darmstadt, Schloßgartenstr. 9, 64289 Darmstadt, Germany

E-mail: `atebartz@ikp.tu-darmstadt.de`

Abstract. Polymer foils with thicknesses 150 nm to around 1 μm were manufactured using the process of spincoating. Varying the concentration of the polymer solution was found to be the most efficient way to control resulting foil thickness. Both interferometric microscopy and energy loss of alpha particles were used for characterization and found to yield agreeing results. For an experiment on laser-driven ion acceleration, these films were used as targets and as substrates for cryogenic hydrogen targets.

1. Introduction

High power laser systems allow for investigation of laser-plasma-interactions and laser-driven ion acceleration for which several mechanisms have been presented in the past. While *Target Normal Sheath Acceleration* (TNSA) [1] takes place in an overdense and thus for the laser impenetrable plasma, laser intensities around $10^{21}\mathrm{W/cm^2}$ enable relativistic transparency for which the ion acceleration mechanism *Laser Breakout Afterburner* (BOA) was presented [2]. This new acceleration mechanism does impose new requirements on both the laser pulse and target characteristics.

In the scope of this work, thin hydrocarbon foil targets with thicknesses in the range of about 150 to 1000 nanometers were manufactured. Equally important to the production process, a reliable method of characterization needed to be found to be able to analyze produced target foils and improve the fabrication process.

2. Choice of material

Due to their uniquely high charge-to-mass-ratio, hydrogen ions, consisting of a single proton, are ideal for acceleration by electromagnetic fields. However, pure hydrogen exists as a gas at room temperature and needs to be cooled below 14 K to form a solid target for laser-driven ion acceleration. Avoiding the inconvenience of cooling to this extremely low temperature, solid hydrocarbons provide proton sources that are easier to handle. Polymethylpentene was chosen for this work for its high hydrogen content with a carbon to hydrogen ratio of 1:2.

3. Manufacturing

Thin foils of solvable materials can be produced using the process of spin coating where the solution is placed on a substrate on a rotatable disk. While the disk is spinned (hence the name of the process), the solution spreads out due to centrifugal force, the solvent evaporates and a thin film of solid remains.

3.1. Preparations

Polymethylpentene is available as a granulate, solvable in cyclohexane at 60°C, which takes about one hour. Preparation of substrates involves application of a separation layer on the substrate as a soluble and thus removable base. This enables foil removal from substrate, to get freestanding targets. Water is a solvent that is easy to handle and doesn't affect the polymer foils, so tensides were used as a water-soluble seperation layer.

3.2. Choice of parameters

We found that the resulting film thickness depends on flow velocity during the spin coating process which can be manipulated by different means. Rotation speed of the spinning disk determines centrifugal force while a change in concentration of the polymer solution changes its viscosity, both affect the flow velocity and thus can be used to influence resulting foil thicknesses. As shown in figure 1, doubling the concentration of the polymer solution more than doubles the resulting film thickness, while doubling the rotation speed only decreases the resulting thickness by a small fraction. Therefore, the concentration was used subsequently as means to control the foil thickness.

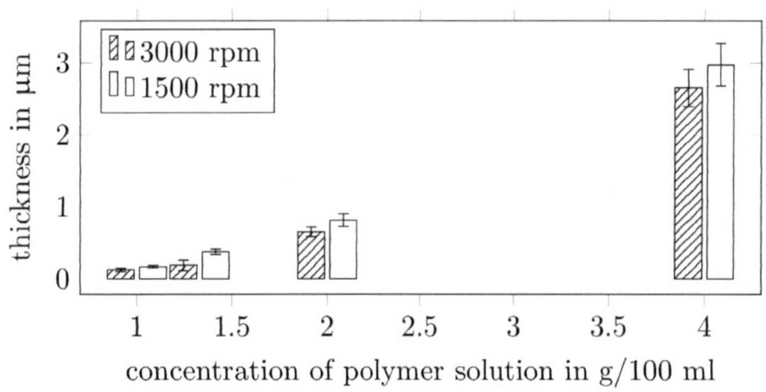

Figure 1. While doubling the rotation speed only decreases foil thicknesses by a small amount, doubling the concentration of the polymer solution more than doubles the resulting foil thickness. The latter has a much bigger influence and is consequently used to control foil thickness.

3.3. Removal from substrate

Target foils are removed from their substrate via dissolution of the separation layer. To achieve this, the substrate is placed in a container at an angle of ca. 45° relative to the horizontal (see figure 2). Then, the water level is raised continuously, thus dissolving the (water-soluble) separation layer and lifting the foil. Now it floats on the surface and can be lifted off. A useful tool for that is a small metal sheet with a hole in it. This creates a target foil with free front and back while the metal part can be glued onto a holder with a geometry tailored to fit the setup for the acceleration experiment.

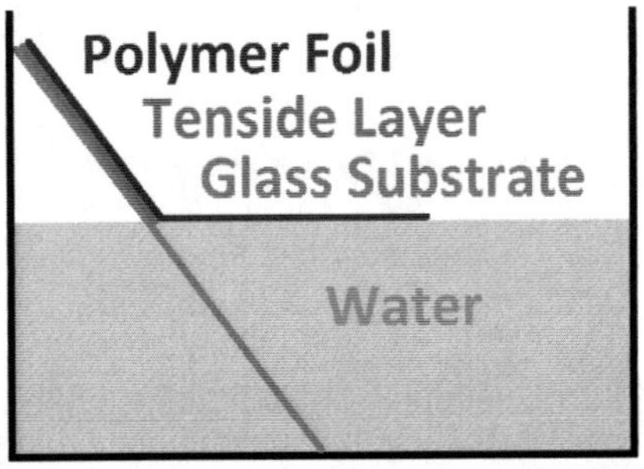

Figure 2. Setup for removal of foils from substrate, together with the order of layers on the glass substrate. A slowly rising water level dissolves the separation layer (middle) and lifts the target foil (top, dark) off its substrate (bottom, grey).

4. Characterization

Being able to measure the thickness and surface characteristics of produced foils is crucial to evaluate their suitability for experiments and adjust and optimize the production process. Two methods were used in the target laboratory of TU Darmstadt: Interferometric microscopy and energy loss of alpha particles. While interferometric measurements characterize the surface, alpha particles traverse the whole volume.

4.1. Interferometric thickness measurements

In our target laboratory, interferometric measurements can be carried out using an optical profilometer. This can be used for quantization of surface roughness and, via measuring the step height between foil surface and substrate surface at the edge of a sample, for thickness measurements. The latter can only be performed on witness samples, not on a foil that was already removed from its substrate and prepared for use in an experiment. Once a foil has been placed on a target holder, it doesn't necessarily lie completely flat anymore which might lead to a step height that is not identical with the thickness anymore. Another drawback of the procedure is that it doesn't give any information on the foil's backside.

An example for a surface profile can be seen in figure 3, corresponding to the lineout presented in figure 4.

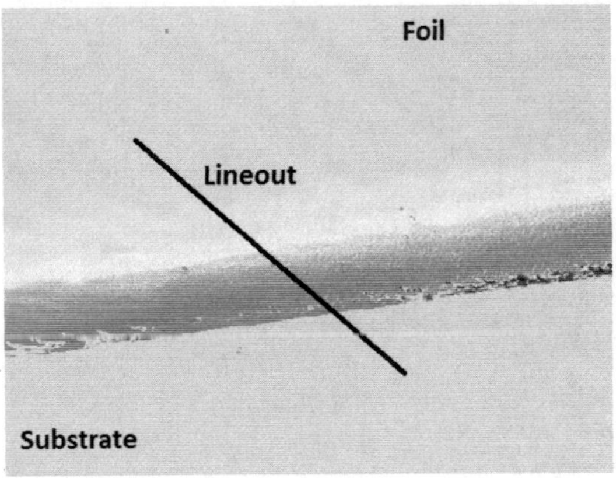

Figure 3. Interferometrically measured surface profile of a foil (upper half) on a substrate (lower half). It is visible that, from creating the edge and thus ripping the polymer foil off on the lower half of the picture, the edge has been disturbed and lifted from the substrate. Since the backside can't be measured, only the increase in height is visible. The black line marks the location of the lineout, displayed in figure 4.

4.2. Energy loss of alpha particles

Alpha particles lose energy while passing through matter mainly through interaction with electrons. With known probabilities for interaction events and energy loss per event, the total energy loss depends on the number of collision partners, thus translating (e.g. via simulation

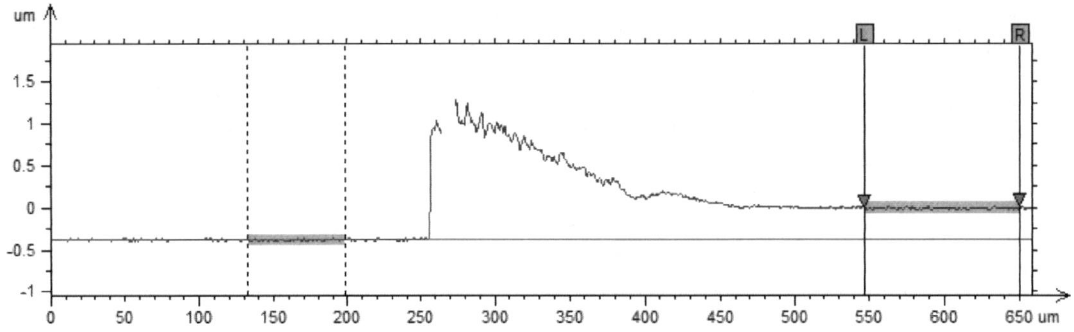

Figure 4. Lineout from the surface profile shown in figure 3. The artificially increased height from the lifted edge is clearly visible. The foil thickness can be determined via the step height, here taken as the height difference between the highlighted, unperturbed sections.

software) to the number of electrons the particle has interacted with. To convert this into thickness information, the material density is required which might be unknown or, in a thin film, different from the density of the bulk material.

Americium-241 was used as a source for alpha particles, it emits them with energies of more than 5 MeV. The energy spectrum is displayed in figure 5, together with the shifted spectrum after passing the sample. The total energy loss of the alpha particles was calculated using the energy shift of the main peak. Measurements of alpha particle energies were conducted using a Schottky diode consisting of gold and silicon. The particles lose their kinetic energy in the detector which creates electron-hole-pairs in the space-charge region. The total charge created corresponds to the amount of energy that was deposited.

A clear spectrum with sharp peaks is needed to be able to measure small energy shifts of the main peak. This requires a thin, thinly coated or uncoated and uncovered radioactive source, as the spectrum shifts and broadens as the alpha particles pass through the source material or coating.

For this work, conversions to thickness were carried out using simulations with the program SRIM (*Stopping and Range of Ions in Matter* [3]) using the density of the bulk polymer given by the manufacturer.

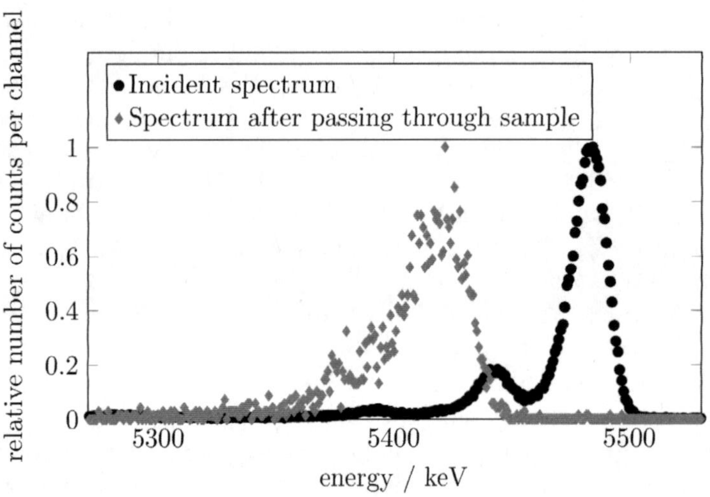

Figure 5. Energy spectrum of alpha particles emitted from Americium-241 (dots) and the shifted energy spectrum (diamonds) after the alpha particles passed through a polymer foil. The energy shift of 63 keV corresponds here to a thickness of 790 nm.

Compared to step height measurements, this method can characterize the exact same sample that is later used for the experiment. However, the measurement yields an average over the whole aperture since the alpha particles are not focused and the source size here was larger than the target size. Consequently, there is neither space-resolved thickness data nor surface information.

The method of thickness measurements via energy loss is promising for characterization of cryogenic targets and has already been utilized for this particular application before. However, due to melting and sublimation, cryogenic targets are not stable and therefore measurements are time-critical. To enhance time resolution, count rates must be increased. Regarding this, the setup can be improved via optimizing source activity, geometry of setup, aperture size and alignment. Considering geometry of setup, special care must be taken to ensure that a large number of alpha particles hits the detector after having traversed the target, while none are detected that didn't pass the sample. While a small degree of spatial resolution is possible with an array of small sources [4] or a mask effectively turning one bigger source into several smaller ones, this severely limits count rates and thus might not be applicable to all setups depending on free aperture.

We propose to describe target foils (of any material) by amount of energy loss rather than thickness. As the total energy loss relates to the number of electrons in the sample, such a description is connected more closely to the actual interaction mechanism between laser and matter which happens via electrons [1, 2]. In addition, the density of the material is not needed for characterization, which might not be provided to sufficient accuracy, depend on properties like chain length or branching in polymers [5] or is only known for the bulk material without insight in how much it might change during manufacture or treatment (coating techniques, sputtering etc. [6]). However, due to the geometry of the characterization process with alpha particles, description via total energy loss is in most cases only suitable for foils with uniform thickness across the aperture.

4.3. Comparison of methods

Results from both methods presented here are displayed in figure 6. Their results for foil thickness are in good agreement. In addition, it is revealed that resulting foil thickness is proportional to the concentration of the polymer solution used for the spincoating process (in the range of 1 to 2.8 g/(100 ml)) which makes thicknesses of new targets predictable.

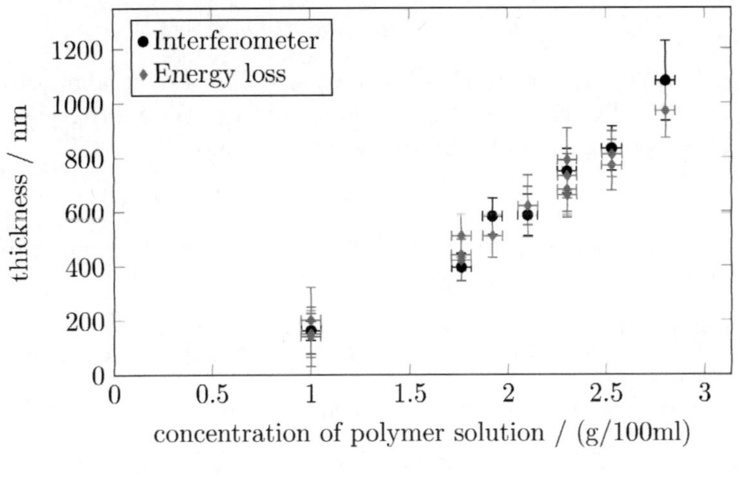

Figure 6. Comparison of interferometrically measured values for target thickness to those calculated from energy loss of alpha particles. A clear proportionality to concentration of the solution is visible, while the thickness measurements from both methods are in good agreement.

5. Experiment

In April 2013, thin foils of polymethylpentene were used as targets in experiments on laser-driven ion acceleration at the *Petawatt Laser for Heavy Ion Experiments* (PHELIX) at *GSI Helmholtzzentrum für Schwerionenforschung* in Darmstadt. The experimental setup included a tilted target to allow for an incident angle of laser direction on target (see figure 7). Some beam profiles exhibit two beam directions as can be seen on a stack of radiochromic film which is colored by ions (figure 8): One beam along target normal which is the expected direction for the acceleration mechanism of TNSA and one along direction of laser propagation which is the suspected ion beam direction for the BOA mechanism. These results indicate that the fabricated target foils were suitable for ion acceleration by not only the TNSA mechanism but also the BOA acceleration scheme (see also [7]).

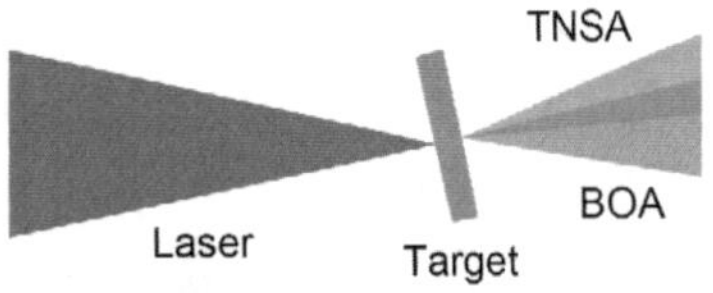

Figure 7. Geometry of experimental setup with expected ion beam directions. The target normal is tilted with respect to laser direction.

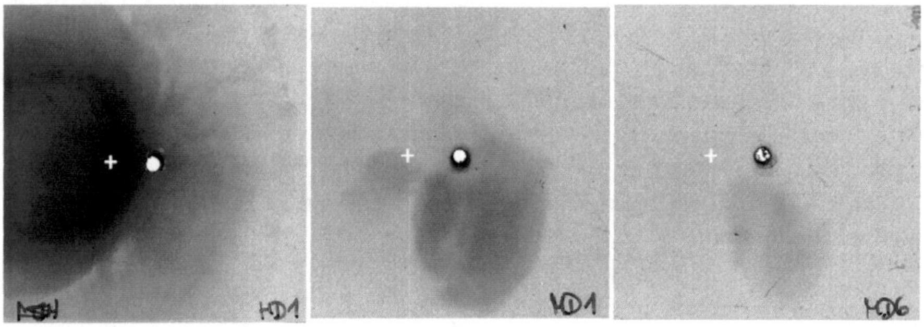

Figure 8. Ion beam profile on radiochromic film, corresponding to ion energies from lower towards higher energies (left to right), exhibiting two distinct beam directions with different maximum energies; one along the target normal (white cross) and one in laser direction (hole in center).

Acknowledgments

We would like to thank Frank Mertins, Tina Ebert and all other members of the target laboratory team, as well as the laser team at PHELIX (GSI), for their dedicated work. We are especially grateful for the financial support by Deutsche Forschungsgemeinschaft (DFG).

References

[1] Wilks S *et al* 2001, Energetic proton generation in ultra-intense lasersolid interaction, *Physics of Plasmas*, **8** 542
[2] Yin L *et al* 2006, GeV laser ion acceleration from ultrathin targets: The laser break-out afterburner, *Laser and Particle Beams*, **24** 291
[3] Ziegler J, Stopping and Range of Ions in Matter (SRIM), Version 2013.00, `http://www.srim.org/`
[4] Fujiwara M C *et al* 1971, Characterization of solidified gas thin film targets via alpha particle energy loss, *Nuclear Instruments and Methods in Physics Research A*, **395** 195
[5] Fetters L J *et al* 1994, Connection between Polymer Molecular Weight, Density, Chain Dimensions, and Melt Viscoelastic Properties, *Macromolecules*, **27** 4639
[6] Anders A 2010, A structure zone diagram including plasma-based deposition and ion etching, *Thin Solid Films*, **518** 4087
[7] Wagner F *et al* 2015, Simultaneous observation of angularly separated laser-driven proton beams accelerated via two different mechanisms, *Physics of Plasmas*, **22** 063110

In-situ formation of solidified hydrogen thin-membrane targets using a pulse tube cryocooler

S Astbury[1], S Bedacht[3], P Brummitt[1], D Carroll[1], R Clarke[1], S Crisp[1], C Hernandez-Gomez[1], P Holligan[1], S Hook[1], J S Merchan[1], D Neely[1,2], A Ortner[3], D Rathbone[1], P Rice[1], G Schaumann[3], G Scott[1], C Spindloe[1], S Spurdle[1], A Tebartz[3], S Tomlinson[1], F Wagner[3], M Borghesi[4], M Roth[3] and M K Tolley[1]

[1]Central Laser Facility, STFC, Rutherford Appleton Laboratory, Harwell Oxford, Didcot, Chilton, Oxon, OX110QX, UK

[2]Department of Physics, SUPA, University of Strathclyde, Glasgow G4 0NG, UK

[3]Institut für Kernphysik, Technische Universität Darmstadt, Schlossgartenstrasse 9, 64289 Darmstadt, Germany

[4]Queen's University Belfast 2015, University Road Belfast, BT7 1NN, Northern Ireland, UK

E-mail: sam.astbury@stfc.ac.uk

Abstract. An account is given of the Central Laser Facility's work to produce a cryogenic hydrogen targetry system using a pulse tube cryocooler. Due to the increasing demand for low Z thin laser targets, CLF (in collaboration with TUD) have been developing a system which allows the production of solid hydrogen membranes by engineering a design which can achieve this remotely; enabling the gas injection, condensation and solidification of hydrogen without compromising the vacuum of the target chamber. A dynamic sealing mechanism was integrated which allows targets to be grown and then remotely exposed to open vacuum for laser interaction. Further research was conducted on the survivability of the cryogenic targets which concluded that a warm gas effect causes temperature spiking when exposing the solidified hydrogen to the outer vacuum. This effect was shown to be mitigated by improving the pumping capacity of the environment and reducing the minimum temperature obtainable on the target mount. This was achieved by developing a two-stage radiation shield encased with superinsulating blanketing; reducing the base temperature from 14 ± 0.5 K to 7.2 ± 0.2 K about the coldhead as well as improving temperature control stability following the installation of a high-performance temperature controller and sensor apparatus. The system was delivered experimentally and in July 2014 the first laser shots were taken upon hydrogen targets in the Vulcan TAP facility.

1. Introduction

Generating ion beams of the desired energy, flux and quality is critical to the advancement of many novel applications[1] such as laser driven ion oncology, Fast Ignition and in fields such as material science. Hadron therapy for example will require nucleon energies $E_{nucleon}$>200 MeV which to date has not been possible to achieve with the current generation of high power lasers.

Experimental work[2] on relativistically-intense ($I > 10^{18}$ W/cm^2) laser systems such as the VULCAN Petawatt laser[3] at the Central Laser Facility (CLF), have facilitated the generation of ion beams of energy 10 MeV > E_{proton} > 75 MeV through the Target Normal Sheath Acceleration[4, 5] (TNSA) method of ion acceleration. TNSA is achieved by irradiating a thin target-foil (on the order of a few microns in thickness) with an ultra intense laser pulse. Energetic fast electrons (MeV-energy) are generated in the laser focus and they are accelerated through the target and propagate out of the rear side of the foil. This escaping charge results in a separation electric field, of TV/m strength which ionizes and accelerates the surface atoms, resulting in the expanding ions travelling quasi normally away from the rear of the target[6]. Any laser pre-pulse of sufficient intensity to ionize the front target surface before the main pulse of the laser arrives, can generate an expanding plasma[7]. This pre-plasma plays a significant role in the subsequent coupling efficiency[8] of the laser and the generation of hot electrons. However, such a pre-pulse can also launch a shock through the target foil[9], and larger target thicknesses must be employed to mitigate damage to the target's integrity (laser contrast dependent). Utilising thicker targets tends to reduce the energy of the resulting proton beam[10] and the intensity scaling ($E_{proton} \propto I^{0.5}$) for TNSA[4,11] alone is not a sufficient driving mechanism to readily meet the needs of applications requiring >200 MeV/nucleon ions.

Simulation studies[12] have suggested the presence of new ion acceleration mechanisms[13] which can be present when an ultra high-contrast laser pulse interacts with a nanometric thick target. By improving the contrast-ratio of the laser pulse, the prepulse plasma is significantly reduced – maintaining target integrity during the early stages of the interaction and enabling further ion acceleration regimes to be accessed, namely enhanced TNSA, Light Sail and Laser Breakout Afterburner (BOA) mechanism. Enhanced TNSA is caused by an improved transfer of the laser energy to the cold target electrons, more efficiently generating hot electrons. These augment the electric fields present in the sheath and cause the ions to accelerate further - particularly in a nanometric target - in a solitary quasi-monoenergetic packet.[14] The Light sail mode utilises radiation pressure[15] to accelerate the ions and the BOA regime manifests itself when the target becomes relativistically transparent. In the latter case in a thin target the electrons are still interacting with the laser field – the propagation of which 'pushes' the electrons, driving a Buneman-instability (a plasmonic instability resulting from a difference in electron and ion velocity of more than the ion-acoustic speed). This imparts a kinetic energy transfer onto the ions at the rear surface of the target.[16] It has been shown via PIC simulations that BOA can theoretically generate ion beams on the order of hundreds of MeV. Such simulations show that the heaviest species present in the target governs the ion velocity of the interaction; thus a low-Z, high purity target is ideal and therefore there is a significant interest in producing solidified pure hydrogen thin films. This paper highlights the work at the Central Laser Facility towards producing an in-situ solidified hydrogen thin film target production capability.

2. Cryogenic target system development:

2.1. Growth Mechanism:

Simulation[17] suggests that for the BOA regime to be realized in solid hydrogen, due to its low-Z, a target thickness of 1-4 μm is ideal. There are three known successful methods of windowless hydrogen target growth, which are: casting[18], extrusion[19] and condensation, all of which were reviewed for efficacy before the development of the system.

The casting method requires the solidified hydrogen to be grown onto a casting plate substrate and then physically removed from a cooled growth surface. Cooling hydrogen using this method can take a long while and due to the stresses involved on the solidified hydrogen during the separation process, thin targets (< 1 mm) cannot reliably be produced and thus use of a casting plate is not a viable method for thin target production.

The second considered approach is the extrusion method. This requires pressurized hydrogen to be forced out of a vessel which has a negative temperature gradient from the top of the vessel to the 'nozzle' out of which the hydrogen is forced as a continuous solidified ribbon. Such a design would require highly pressurized hydrogen inside a vacuum chamber which may not be suitable due to risk of debris damage. Spatial instability is also an inherent issue in such designs as the hydrogen exits the nozzle and as spatial tolerance is minimal for the focus of the beam, this method is unsuitable for the application discussed.

After considering the above alternative approaches, the condensation method was decided upon for several reasons: lower hydrogen gas pressures are preferential in the laser interaction chamber, quick target growth after system-cooldown and high spatial stability of the hydrogen target.

Hydrogen growth by condensation involves injecting gaseous hydrogen into a sealed environment, henceforth referred to as the growth chamber, in which is a cold surface. This surface is then cooled below the boiling point of H_2 causing the hydrogen gas to condense and fill a target substrate which is then further cooled below the freezing point (T~13.9 K) to solidify the liquid into a thin H_2-ice layer. However, due to the requirement of having a pure hydrogen target, a windowless substrate-free target was an essential criterion of the system.

2.2. Design Parameters:

The main design parameters of the system involve reaching a sufficiently low temperature using a cryogenic cooling system and control of the environmental conditions in the target growth chamber. Figure 1 shows the phase diagram of hydrogen at low pressures, which illustrates the pressure:temperature relationship of hydrogen and its phase transitions.

In order for successful condensation and freezing of a thin-film of hydrogen, a relatively constant vapour pressure on the order of several hundred mbar must be maintained as well as fine temperature control between 10 – 25 K, with a resolution of ~0.3 K; thus a sensor/heater feedback-loop temperature control system was required. From Figure 1, it can be seen that achieving liquefaction requires a minimum pressure of 70.4mbar to be maintained within the growth environment. As this is surrounded by a high vacuum, a leak-tight seal was mandatory and to this end a pressure sensor/readout was critical.

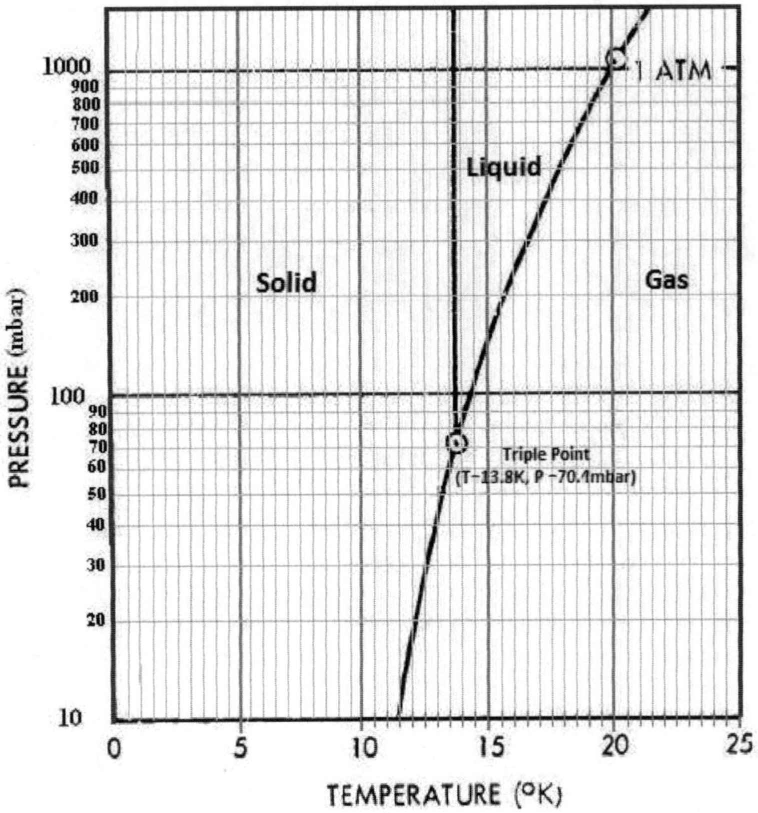

Figure 1. Low pressure phase diagram of hydrogen, showing the solid, liquid and gas regimes and their dependence on pressure and temperature.[20](Reduction from reference)

A hydrogen gas inlet and pump system was necessary to both inject and evacuate the growth chamber during the growth process as well as a sufficiently sized outer vacuum chamber to house the system along with access flanges for a turbo pump, various feedthrough ports and optical diagnostics.

The heater/control system served a secondary purpose to decrease the time taken warming the system back to room temperature before venting the vacuum chamber. This was necessary to prevent condensation/freezing of contaminants onto the apparatus between cooldown/warm-up cycles.

2.3. Conceptual Design:

After consideration and research of the design parameters, a prototype system was designed and installed at CLF[21]. This comprised a Sumitomo RP-082B pulse tube cryocooler which provided 1 W of cooling power at 4.2 K via helium gas compression refrigeration. Such a system does not require cryogenic liquids for cooling, negating the impracticality and cost of cooling via liquid helium.

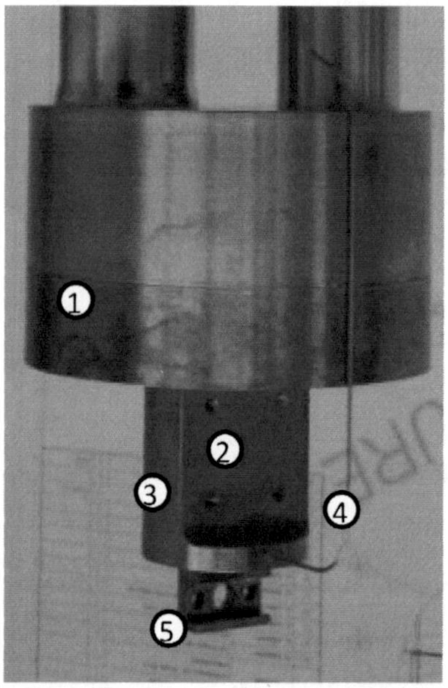

Figure 2. Cryocooler copper coldhead and target mount adapter showing the relevant components of the coldhead.

Figure 3. Lower section of the assembly showing the target mount and growth chamber (1), radiation shielding (2) and pressure transducer (3).

The thermal properties of the system were controlled by two 25 Ω resistor heaters and two platinum RTD temperature sensors (mounted at figure 2, label 2 and figure 2, label 3 respectively). Temperature was regulated in the form of an Omega CYC325 temperature controller. This system received input data from the calibrated temperature sensors and controlled the power to the heaters to maintain a setpoint temperature governed by a PID control feedback loop. The total heater output power for the controller was 27 W (1 x 25 W, 1 x 2 W).

The coldhead of the system (figure 2-1) was machined from copper and was designed to provide a high thermal capacity for the target mount to limit susceptibility to sudden temperature spikes which would quickly evaporate the target due to the volatility of hydrogen.

The hydrogen inlet capillary (figure 2-4) also functions as the evacuation line connected to a roughing pump line. This was designed to precool the gas by passing it through the coldhead before injection into the growth chamber and was manually operated using a series of hand valves.

The growth chamber (figure 3-1) comprises a cylindrical polymethyl methacrylate (PMMA) 'cup' which was bolted to the base of the coldhead. A dual indium ring seal ensured the cup was leak-tight. Pressure inside the growth chamber was measured using a transducer (figure 3-3) which connected to the growth chamber via a stainless steel capillary and displayed a digital pressure readout in real-time.

A 2mm thick aluminium shroud acted as a radiation shield (figure 3-2) to prevent thermal radiation from any external heat loads that could raise the temperature around the target. The copper target mount (figure 2-5) consisted of a copper shaft subtending around 25 mm from the coldhead which a target foil could be mounted upon.

The target foils (figure 4) were necessary to promote condensation and growth of the target over an aperture in its centre. These foils were designed to be easily replaceable after the laser interaction with the hydrogen target which would cause inevitable damage to the foil and were positioned horizontally to provide as much thermal contact to the coldhead as possible. Several designs were machined for future research into manipulating the flow of liquefied hydrogen to saturate the target aperture, however for the scope of this project a circular aperture sufficed.

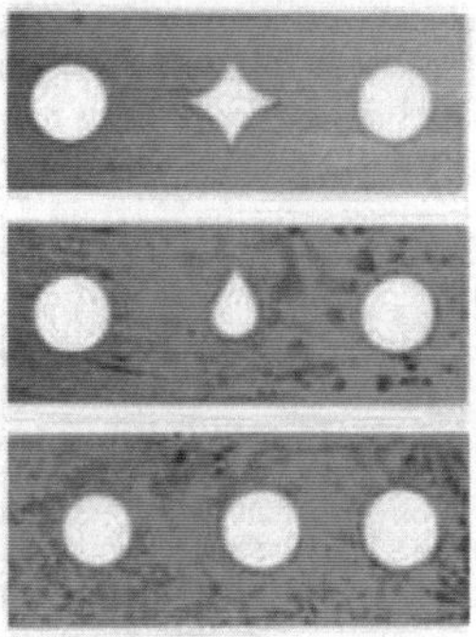

Figure 4. Horizontally mounted copper target foils of various central aperture shapes; spiderweb (top), teardrop (centre) and circular (bottom)

An outer vacuum chamber was constructed to house the system which had a port for a roughing and turbo pump to allow for a vacuum on the order of 1×10^{-5} mbar along with feedthrough ports for the components and diagnostics discussed above.

2.4. Base temperature minimisation:

After constructing and assembling a working system within a leak-tight vacuum chamber, research began on achieving a substantially cold base temperature to allow for freezing and survivability of the hydrogen target, i.e. a maximum of 13.9 K (approximate freezing point of hydrogen through the liquid phase at $P > 70.4$ mbar).

By running the pulse tube with the heaters off, the results in Table 1 were gathered which shows the effect on base temperature with and without radiation shielding and superinsulation blanketing. Note for this testing that no hydrogen was injected into the growth chamber which was instead pumped, thus pressure $P \rightarrow 0$ (limited by the impedance of the 1mm capillary from the roughing pump through which it was evacuated). Also worth noting is the fact that the pulse tube cryocooler operates in two stages; 40 W cooling power at 45 K on the first, upper stage and 1 W cooling power at 4.2 K on the second, lower stage.

Journal of Physics: Conference Series **713** (2016) 012006 doi:10.1088/1742-6596/713/1/012006

Shielding	Minimum Temperature (K)
No shielding	14 K ± 0.5 K
Aluminium one stage radiation shield (figure 3-2)	8.4 K ± 0.3 K
Radiation shield and superinsulation blanket	5.8 K ± 0.3 K

Table 1. Base temperature achievable upon varying the shielding surrounding the coldhead.

As Table 1 shows, running the pulse tube without any surrounding radiation shielding (with the relevant diagnostics/components attached) meant that the core temperature could not fall below 14 K. Due to the requirement of $T<13.9$ K to achieve solidification, the system in this state was not suitable for target growth and thus radiation shielding on the first stage was developed. This effectively meant that a 40 K shielding (discounting inherent negative temperature gradient downwards along its surface) surrounded the coldhead, limiting the effects of thermal radiation and dropping the core temperature to a minimum of 8.4 K. Finally, a superinsulation blanket was installed surrounding the radiation shield; increasing the cooling to a minimum temperature of 5.8 K. Thus an 8 K tolerance was achieved between base temperature and the melting point of hydrogen to mitigate the rate of sublimation/target thinning and research into establishing a repeatable growth procedure was commenced.

2.5. Growth Procedure Research:

Upon achieving a suitably low base temperature, work began on establishing a repeatable growth procedure for solidified hydrogen targets in situ. The stability of the temperature sensors between ~6 and 25 K was ± 0.3 K and due to the manual gas injection and readout of the display, pressure readings were accurate to ±20mbar.

As the system is based on the condensation method of hydrogen solidification, referring to Figure 1 it is apparent that the pressure of hydrogen must be above 70.4 mbar in order to pass through the gas:liquid phase boundary.

Once in the liquid regime, the coldhead is further cooled and the vapour pressure maintained by inletting a steady supply of hydrogen into the target chamber due to a pressure decrease resulting from the liquefaction of the gaseous molecules. However, achieving this with this particular setup was difficult as the injection system consisted of a series of manual hand valves and could not be self-regulated. When the temperature of the coldhead passes below ~14 K, nucleation sites within the liquid are observed which initiates the phase transition from liquid:solid and a crystalline solid hydrogen structure then forms as the target freezes.

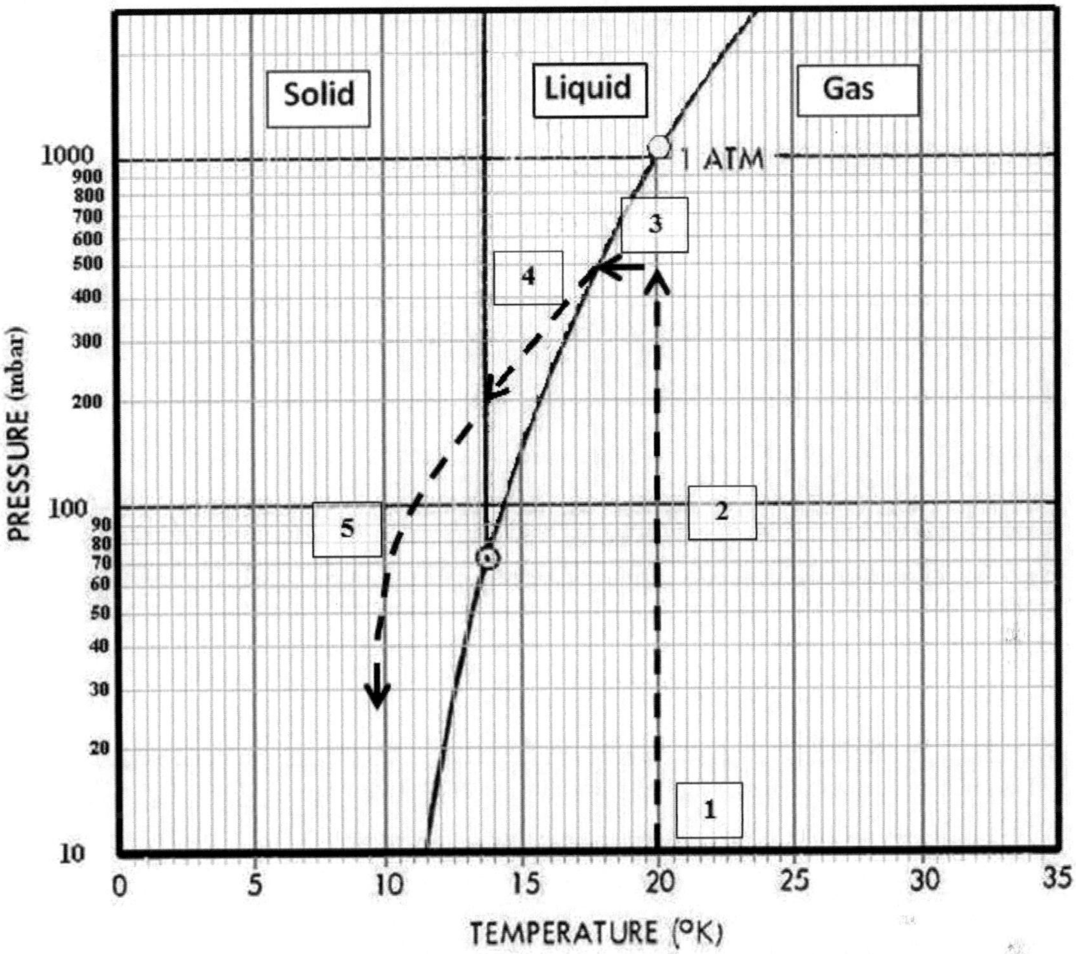

Figure 5. Illustrated phase diagram showing the growth procedure which was devised and followed for in-situ target production.

Referring to Figure 5, there are five relevant stages within the followed growth procedure:

1. The initial state of the system: Regulating the temperature through a temperature controller at 20 ± 0.3 K (above the boiling point of hydrogen).
2. Hydrogen gas injection: The growth chamber is slowly filled with hydrogen until the pressure reaches ~500 ± 20 mbar. This ensures there is a significant pressure to ensure the hydrogen can enter liquid phase as liquefaction causes an inherent pressure decrease.
3. Once there is an appreciable pressure inside the growth chamber, the temperature is dropped and a steady inflow of gas is injected into the chamber to alleviate the pressure drop due to the fall in temperature. When the gas:liquid phase boundary is passed, droplets of hydrogen liquid can be observed flowing down the coldhead/target mount.
4. As liquefaction occurs, there is a sudden pressure decrease within the chamber. Liquid hydrogen begins to flow/drip down the coldhead and over the target aperture: the colder the temperature falls, the lower the vapour pressure and the more viscous the liquid H_2 becomes.
5. When the liquid:solid phase boundary is reached (~14 ± 0.3 K), dependent on the rate of cooling, one or more nucleation sites can be seen on the liquid. From these points, a crystalline structure of solidified hydrogen begins to form, eventually freezing into a thin layer over the target aperture.

Journal of Physics: Conference Series **713** (2016) 012006 doi:10.1088/1742-6596/713/1/012006

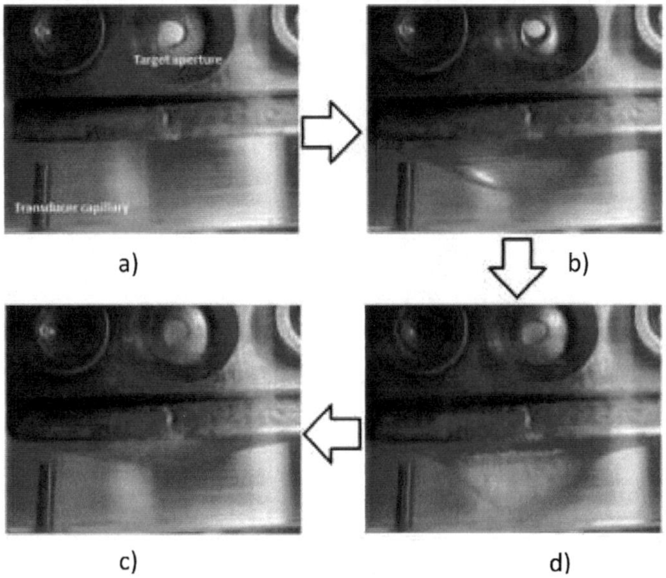

Figure 6. Time-lapse of the growth process. a) Initial state of system at 20 K (stage 1). b) Liquid-phase (stage 4). Note the droplet which subtends from the bottom of the coldhead and the liquid ring over the target aperture. c) Freezing of the liquid film (stage 5). d) Solid H_2 slowly sublimes due to low vacuum conditions within growth chamber. (Images taken with a USB-powered microscope).

The main difficulties realized with this particular system design were:

- *Liquid saturation over the target aperture*: The embedded nature of the target foil – which was sandwiched within the target mount – meant that 'catching' the liquid over the target centre was temperamental and often would take multiple attempts. This was to be revised along with several other design modifications.
- *Fixed/bolted growth chamber*: As the initial configuration was designed as a proof-of-concept, the growth chamber was fixed in place. The ultimate focus of the project was to research the laser interaction with the hydrogen target and as such it is essential that the growth chamber be removable in-situ by designing a removable sealing mechanism.
- *Inaccuracies and non-repeatability of gas injection:* As discussed, the manual nature of the gas injection mechanism led to the fact that maintaining a specific pressure within the growth chamber was difficult. The fluctuations in pressure due to the high sampling rate of the transducer lead to an appreciable error in pressure readout from the display. This could be alleviated by installing data logging hardware/software.

3. Design Modifications

3.1. Dynamic sealing mechanism:

In order to address the experimental issues with the initial system design, several design modifications were made. The primary concern was to design and implement a dynamic sealing mechanism to remove the growth chamber from the coldhead, expose the target to the laser, and reseal the growth chamber, all operated remotely within the control room of the laser facility.

Initial research began on a sealing mechanism which was compatible within the laser chamber i.e. operable at cryogenic temperatures and would enable removal of the the growth chamber from the proximity of the laser before interaction and would not introduce compressed gases/gas lines nearby. An O-ring seal and a gate valve were not suitable for the application for these respective reasons.

After researching various sealing designs, it was deemed necessary that a motor-drive system would be ideal for the experiment allowing growth chamber withdrawal and re-seal by driving the motor downwards and upwards, respectively. This required that the growth chamber be forced against some sealant material that is reusable and suitable to withstanding the large pressure gradient between the inside of the growth chamber (~100 mbar) and the outer vacuum chamber (10^{-5} mbar).

3.1.1. Sealant material:

The design focussed on placing the sealant material either on the cold head or the rim of the growth chamber and the motor forcing the growth chamber against the coldhead with the sealant material in-between. It was essential that this material should be easily replaceable and adequate for multiple sealings.

Indium pads have been used for similar applications in the past; however this was impractical for this particular setup for two reasons. Firstly, the indium pads are difficult to remove due to the moulding of the metal around the coldhead. Secondly, and importantly, the frequency of reseals would mean regular replacements which are both expensive and time-consuming, which is not ideal during the laser experimental beam-time.

To address the above issues, a novel sealing mechanism, as shown in figure 7, was designed and tested for efficacy. Two and three-bladed knife-edge adapters were machined, to be attached to the rim of the growth chamber which would be forced into a polyimide (Kapton) film of sufficient thickness. The motor drive was initially replicated by a mechanism whereby a threaded screw was centrally aligned on the underside of the growth chamber and set at various torques and forced upwards into a ring of Kapton. This mechanism was then connected to a helium mass spectrometer monitor whereby the area inside the growth chamber is pumped and evacuated and helium gas is injected around the seal. Any helium particles that penetrate the seal are detected by the spectrometer giving a leak-rate; the acceptable value of which for a gas tight seal is < $1\text{x}10^{-7}$ mbar·l/sec.[22]

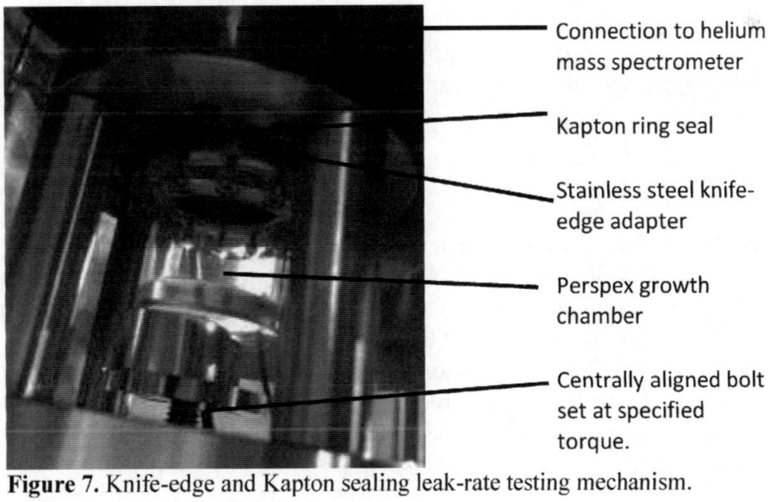

Figure 7. Knife-edge and Kapton sealing leak-rate testing mechanism.

By adjusting the torque on the bolt on the underside of the growth chamber, the force acting upon the Kapton from the knife-edges could be varied; too little torque and the blades did not exert enough force on to the disc to provide a gas-tight seal and too much torque resulted in the blades penetrating the rear of the Kapton again, breaking the seal. The results of the leak testing with two different thicknesses of Kapton with a two-blade and three-blade adapter are shown in figure 8:

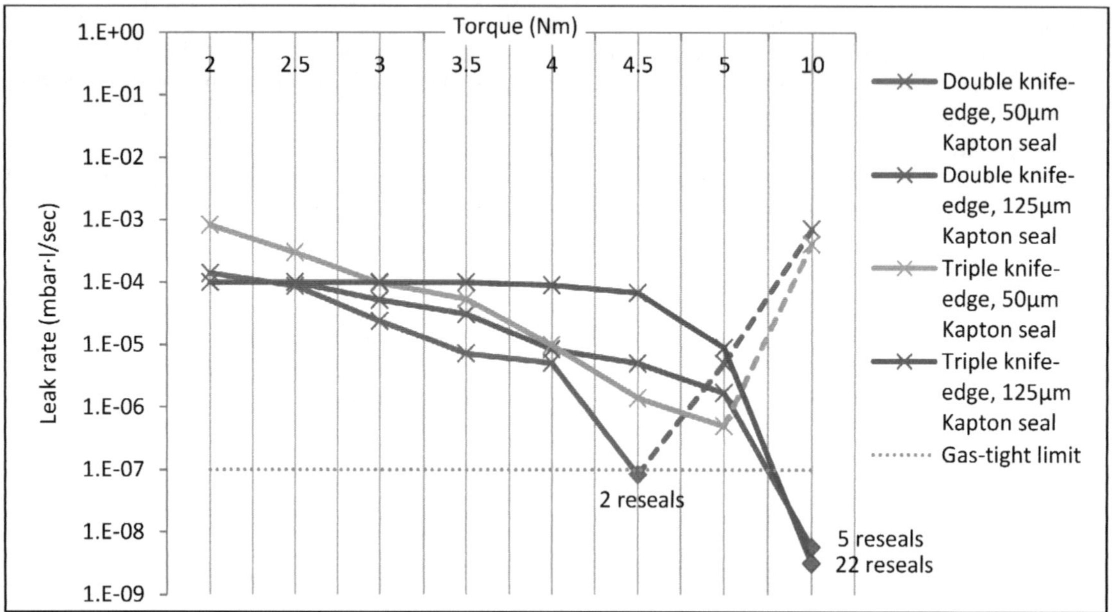

Figure 8. Line graph depicting the leak rates of both a double-edged and triple-edged knife adapter upon different thicknesses of Kapton film (50 µm and 125 µm) when various torques were applied to a bolt forcing the knife-edges into the film. For successful gas-tight seals, the number of repeatable seals are shown.

From the results shown in figure 8, there were only 3 gas-tight configurations of the system. Of the three successful tests, a triple-knife edge adapter under 10Nm of torque forced upon 125 µm of Kapton provided the most leak tight seal of $3.2\text{x}10^{-9}$ mbar·l/sec and was repeatable 22 times at room temperature before the Kapton film failed and needed replacing. Such a result was adequate for use within the laser chamber, meaning that one film installation could last multiple shots. 50 µm films provided better leak rates than the 125 µm films at the same given torque as they were found to be more yielding. However, as illustrated by the dashed lines on the graph, at higher torques the films were cut through completely, leading to a rapid deterioration in the leak rate. For this reason, it was decided that a triple knife-edge adapter against a 125 µm film with a driving torque of 10 Nm would be used for the dynamic sealing mechanism.

3.1.2. Motor drive:

After having established the relevant parameters for achieving and maintaining a gas-tight seal, a remotely controlled motor mechanism was required to allow for the removal/closing of the growth chamber within the laser chamber. This had to be designed to deliver 10 Nm of torque, as well as being sufficiently insulated to avoid undesirable heat loading to the coldhead.

The resulting design was a stepper motor to which the growth chamber was fastened, enabling translational motion and delivering appreciable force onto the Kapton to achieve a gas-tight seal and withdrawal of the mechanism to allow laser access to the hydrogen target after growth, operated from a remote control logic box.

3.1.3. Growth chamber redesign:

Due to the focal length of the optical diagnostics required for monitoring the target, regarding which the author refers to the work of Schaumann G[23], there was a requirement to design a narrower growth chamber. Ultimately, this design resulted in a change in shape of the growth chamber – from a circular to a rectangular footprint which in turn necessitated narrower wall thickness. The knife-edge adapter was also redesigned to match the dimensions of the new design and rather than attached to the rim of the growth chamber due to size constraints was instead bolted onto the base of the coldhead.

After having machined a narrow chamber, integrity testing with the motor revealed cracking in the structure of the perspex material and it was deemed necessary that the walls of the growth chamber required a tougher material which would not be susceptible to fracturing which still allowed visibility of the target. Further testing warranted another redesign which led to the final design of the growth chamber being constructed from stainless steel walls with two quartz optical windows in line with target centre to allow in-situ target viewing. The entire driving mechanism was connected to the coldhead with thermally insulating glass fibre rods.

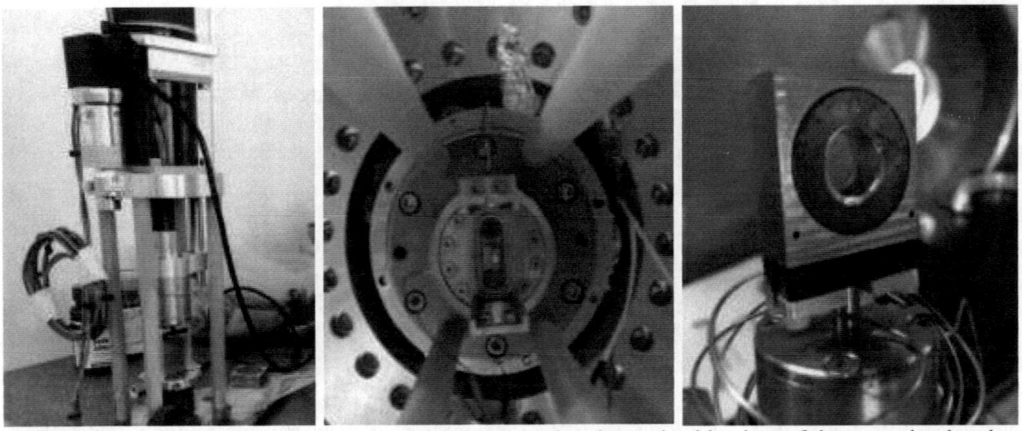

Figure 9: (Left to right) a) 10Nm motor driving mechanism b) Underside view of the mounting brackets for the motor along with the knife-edges. c) Stainless steel growth chamber with transducer capillary.

It is worth noting, however, that due to the large thermal mass in the form of the motor system attached to the base of the coldhead the baseline temperature achievable by the system rose to 9.8 ± 0.3 K with the growth chamber driven into the Kapton and 8.5 ± 0.2 K when the growth chamber was retracted (with superinsulation wrapping in both instances). The change of the material of the growth chamber from an insulating perspex to a conductive stainless steel also added to this rise in temperature. Despite this, the base temperature was still well within tolerance to establish freezing of the liquid hydrogen.

3.2. Gas flow system upgrade:

An automated gas-flow system was commissioned which facilitated the control of the gas pressure inside the growth chamber by using a series of reservoirs and mass flow controllers. This allowed regulation of the flow of hydrogen into and the pumping rate out of the growth chamber in order to maintain a set-point pressure which is crucial in the growth process. This comprised a compact Reconfigurable Input/Output (cRIO) controller which ran through a LabVIEW VI to regulate all aspects of the growth process from a single interface.

The LabVIEW program was modified by Alex Ortner, *Technische Universität Darmstadt*, to incorporate the features necessary for the eventual installation of the experiment into the laser target chamber which included: gas injection/pumping, motor driving, temperature control, data logging and camera viewing. This also meant that human errors were negated from the readout of temperature/pressure values which often fluctuated due to the nature of the system allowing for a more accurate and repeatable growth process.

Journal of Physics: Conference Series **713** (2016) 012006 doi:10.1088/1742-6596/713/1/012006

Figure 10: Automated gas-flow system allowing for gas pressure regulation of the growth chamber.

3.3. Temperature control and heater upgrades:

Alongside the integration of the automated gas-flow system an upgrade was necessary to the temperature controller which could only read two temperature sensors simultaneously and output just 27 W of heater power. A model upgrade to the Lakeshore 336 system allowed the monitoring of 4 concurrent sensors and provided 150 W of heater power output. This also allowed for improved PID feedback which improved the temperature regulation of the heaters.

Alternatives were researched in order to improve the warm-up time – roughly 90 minutes with the initial setup. A 100 W and 50 W cartridge heaters were decided upon to both utilise the whole power output of the new temperature controller and to offer volumetric heating as opposed to surface heating. This modification improved warm-up time by a factor of 3 which was crucial to ultimately allow for a higher shot rate.

A further modification was made to the temperature sensors. Initially these were platinum surface mounted sensors which were very susceptible to the effects of thermal radiation and were also difficult to replace which would be necessary due to the potential of EMP damage from the laser pulse. Modified PT-100 temperature sensors were designed and calibrated at Technische Universität Darmstadt which allowed for embedded mounting within the coldhead, leading to a more stable temperature readout from ± 0.3 K to ± 0.1 K (along with an improvement in radiation shielding) at $T < 20$ K. The sensors were attached to replaceable cryogenically compatible connectors to allow for fast in-situ replacement of damaged components.

3.4. Target mount redesign:

Whilst researching a repeatable growth cycle method, there was an intermittent issue in that when the hydrogen gas began to condense, the liquid flow would not always form over the aperture in the centre of the target foil and instead would often trail down the edge of the foil and drip onto the base of the warmer growth chamber, evaporating back into gaseous phase and re-condensing in the same region.

By redesigning the target mount, the target foil was mounted vertically (Figure 11) in order to manipulate the liquid flow down the foil and saturating the aperture in a repeatable manner. After evaluating the cause of the issues with the original target mount/foil (Figure 4), it was found that the graining of the foil was vertical, meaning that liquid flow would follow these contours due to the low viscosity of liquid hydrogen. Owing to the nature of laser cut foils, there was a slight raised edge around the central aperture of the foil forcing the liquid flow around the outer edge preventing saturation over the aperture. By having the foils machined such that the graining was horizontal in the new design, the liquid was allowed to flow more evenly down the width of the foil and consistently saturate the target aperture.

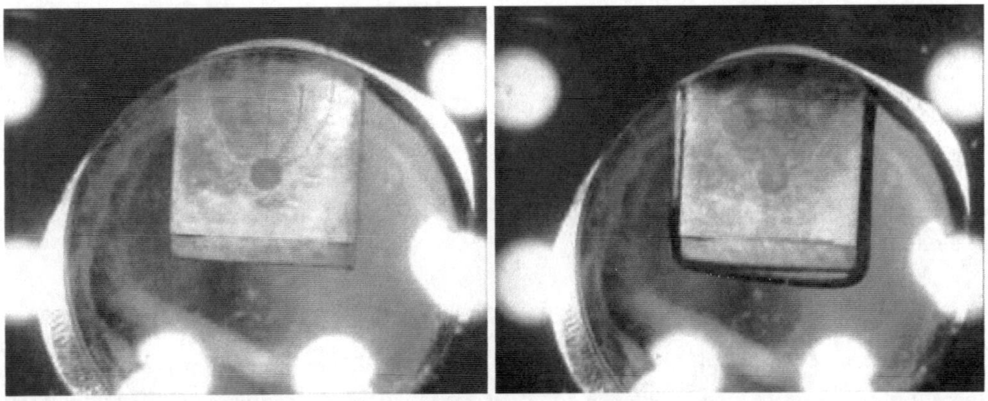

Figure 11: Images of the improved target foils and orientation redesign prior to (left) and after (right) target growth. The foils above have grooves leading towards the centre to promote liquid flow over the target aperture.

3.5. Two-stage radiation shield:

A design adjustment for the existing radiation shield was deemed necessary to further reduce the base temperature of the coldhead. Designing a secondary inner radiation shield connected to the colder (4.2 K) second stage of the pulse tube meant that any radiative heat from the inside of the outer shield would be reflected by the cooler inner shield as opposed to heating the target mount. The design was altered to facilitate the replacement of target foils and temperature sensors by being machined in 2 halves such that the front half could be removed exposing the bolts connecting the target foil to the mount as well as providing access to the temperature sensor connectors.

By integrating the two-stage shield onto the system, a base temperature drop was achieved on the order of ~1.3 K, thereby reducing the minimum temperature achieved by the single stage shield (with superinsulation blanket) from 9.8 ± 0.3 K to 8.5 ± 0.1 K with the growth chamber closed and from 8.5 ± 0.2 K to 7.2 ± 0.2 K with the chamber retracted.

4. Experimental Delivery

In July 2014, the cryogenic targetry system was installed into TAP for an experiment investigating proton acceleration via the BOA mechanism from thin hydrogen targets. By suspending the system from a motorized XYZ stage within a re-entrant chamber on the roof of the target chamber, the position of the target foil could be aligned with the beam axis.

The system was operated in situ, allowing for the first cryogenic target interactions in the Vulcan Petawatt facility since its inception. At the time of writing, data analysis of the results from this experiment is ongoing.

During experimental beamtime it was found that upon retracting the growth chamber to expose the target, the solid hydrogen quickly heated up and sublimed, severely limiting target survival time. Research into this issue led to the conclusion that it was caused by a warm gas effect whereby residual hydrogen ice on the inner edge of (and gaseous hydrogen within) the growth chamber rapidly heated upon losing contact with the coldhead. The escaped gas undergoes heat exchange with the warmer heat loads within the target chamber and is either pumped out or cryo-pumped back onto the coldhead. This latter effect is evidenced by characteristic temperature spikes on the sensors and causes rapid heating of the target. As this is a surface-heating effect, the temperature spike typically lasts for around

30 seconds, rising by approximately 10 K in the original test chamber. This effect had to be accounted for when activating the laser pulse to ensure the target had not sublimed before interaction.

5. Future work and areas of research

As discussed above, an issue of temperature spiking manifested upon retraction of the growth chamber caused by a warm gas effect. There are two methods of assuaging this issue: evacuating the gas to limit the heat loading upon the target from its re-condensation and lowering the achievable base temperature of the target mount to prevent the temperature from rising above the melting point as a result of the temperature spike.

5.1. Mitigation of thermal loading:

To minimise heat loss at the bottom of the target mount due to the temperature gradient stemming from the bottom of the pulse tube, a redesigned target mount adapter is being developed which reduces the length compared to the original adapter (Figure 2).

In order to better understand the heating mechanisms about the target focus has begun on thermal profiling of the system. This required an improvement in the temperature reading accuracy by installing high-precision Cernox™ sensors. Improvements are also being made to the radiation shield and higher quality superinsulation has also been incorporated which will help to further reduce the thermal load.

5.2. Improving pumping capacity:

Experimental observations have shown that the pumping rate is a significant factor in reducing the warm gas effect from heating the target mount; a greater pumping capacity resulting in reduced temperature spiking. Having a larger vacuum chamber volume also alleviates the issue due to greater dispersion of the gas.

A new testing chamber has been built within the CLF to accommodate such modifications which houses two roughing pumps and two turbo pumps with a combined pumping capacity in excess of 2000 l/min, increasing the capacity from the original test chamber by a factor of 4 and improving the vacuum by two orders magnitude.

Figure 12: CLF's improved cryo-targetry test chamber.

5.3. Obtaining thinner target films:

After having improved the cooling efficiency and the vacuum of the test system, achieving increasingly thin target films which is preferential in laser-interaction experiments is to be focussed on. Due to issues with diagnostics in measuring and controlling target thickness at the time of writing, the hydrogen films produced by the system are estimated to be comparable to the target foil thickness – on the order of a few hundred microns.

There may be several routes which can successfully achieve this which are intended to be explored, for instance mounting a comparably thin target foil, and target growth via deposition as opposed to the condensation-freezing method, which is possible following the integration of the automated gas-flow system. Reducing the thermal loading upon the coldhead as well as mitigation of the warm-gas effect will be beneficial to this end.

5.4. Target thickness-measurement diagnostics:

In situ thickness measurement has been a key area of research for utilising the system in laser-target interaction experiments and firing upon a target which is too thick will not generate ions of sufficient energy. Due to the volatile nature of hydrogen it is crucial that the thickness of the solid hydrogen film is characterised in real-time to account for the sublimation rate of the target when exposed to the high vacuum and comparatively high thermal loading of the target chamber.

Two methods for target thickness monitoring have been explored: α-source measurement and a confocal chromatic sensor. Obtaining a strong enough signal from the α-source over the working distance within the chamber proved cumbersome and so research in this domain is being focussed on a confocal chromatic sensor system.

5.5. Thermal modelling:

Establishing a clear understanding and simulated model of the thermal effects about the target centre will be crucial to refer to prior to design modifications. COMSOL modelling of the thermodynamic effects due to black-body radiation, convective heating inside the target chamber due to 'warm' gas flow upon hydrogen injection and enthalpic contributions due to phase transitioning will be a necessary investigation.

Acknowledgements

The authors gratefully acknowledge funding from EPSRC grant number EP/K022415/1 and in particular to Prof Marco Borghesi for supporting the project.

The authors also gratefully acknowledge funding from Laserlab-Europe grant no. 284464.

CLF would like to extend their gratitude to Gabriel Schaumann, Stefan Bedacht, Alex Ortner (especially for his contribution to the LabVIEW programming), Florian Wagner and Alexandra Tebartz of Technische Universität Darmstadt for their collaboration.

Data associated with research published in this paper can be accessed by contacting the corresponding author.

References

[1] H. Daido et al, Review of laser-driven ion sources and their applications Rep. Prog. Phys. 75, 056401, 2012

[2] Roth M and Schollmeier M 2013 Ion Acceleration: TNSA *Laser-Plasma Interactions and Applications* ed. McKenna P, Neely D, Bingham R and Jaroszynski D chapter 12 pp 329-332

[3] C. Hernandez-Gomez et al, Vulcan petawatt-operation and development, *J. Phys. IV France* **133**, 555-559, 2006

[4] L. Robson et al, Scaling of proton acceleration driven by petawatt-laser-plasma interactions, *Nat. Phys.* **3**, 1, 58-62, 2007

[5] S. C. Wilks et al, Energetic proton generation in ultra-intense laser–solid interactions, *Phys. Plasmas **8***, 542, 2001

[6] J. Green et al, Enhanced proton beam collimation in the ultra-intense short pulse regime, *Plasma Physics and Controlled Fusion*, **56**, 084001, 2014

[7] A. J. Mackinnon et al, Effect of plasma scale length on multi-MeV proton production by intense laser pulses, *Phys. Rev. Lett,* **86**, 9, 1769-1772, 2001

[8] R. Gray at al, Laser pulse propagation and enhanced energy coupling to fast electrons in dense plasma gradients, *New J. Phys*. 16, 113075, 2014

[9] D. Carroll et al, Dynamic control and enhancement of Laser accelerated protons, *Comp. Rendus Physique* **10**, 2-3, 188-196, 2009

[10] D Neely et al, Enhanced proton beams from ultra-thin targets driven by high contrast laser pulses, *Appl. Phys. Lett*. **89**, 2, 2006

[11] C. Brenner et al, Modelling the effect of laser focal spot size on sheath-accelerated protons in intense laser-foil interactions, *Plasma Physics and Controlled Fusion*, 56, 084003, 2014,

[12] Albright B J, Yin L, Bowers K J, Hegelich B M, Flippo K A, Kwan T J T, and Fernández J C 2007 *Phys. Plasmas* **14** 094502-2

[13] M. Borghesi et al, Laser-driven ion acceleration State of the art and emerging mechanisms Nuclear Instruments & methods in physics research A, 740, 6-9, 2014

[14] Yin L, Albright B J, Hegelich B M and Fernández J C *Laser and Particle Beams* **24** pp 291-298, 2006

[15] B. Quai et al, Stable GeV Ion-Beam Acceleration from Thin Foils by Circularly Polarized Laser Pulses, Phys. Rev. Lett 102, 14, 145002, 2009

[16] Roth M, Bedacht S, Busold S, Deppert O, Schaumann G, Wagner F, Tebartz A. and Schollmeier M, Breaking the 70MeV proton energy threshold in laser-proton acceleration and guiding beams to applications. *IPAC2014 conference 15-20 June, 2014*

[17] B. M. Hegelich et al, Laser-driven ion acceleration from relativistically transparent nanotargets, *New Journal of Physics* **15,** 2013

[18] Ishimoto S, Kobayashi T, Morimoto K., Nomura I, Ozawa A, Suzuki S, ... and Tsuru T 2002 *Nuclear Instruments and Methods in Physics Research A* **480** pp 304-314

[19] Markov A N, Fradkov A B and Chemetskii V D 1977 *Sov. J. Quantum Electron.* **7** p 641

[20] Keller C W *et al* 1967 *Handbook of Physical and Thermal Property Data for Hydrogen* p 3.4

[21] Tomlinson S T, Spindloe C, Holligan P and Tolley M 2012 Design for Production of Thin Film Solid Hydrogen Targets *Central Laser Facility Annual Report 2012-2013* p 2

[22] Theoretical Basics and Applications of Helium-Leak Detection 2007 *Pfeiffer Vacuum* http://www.redcross.org.uk/index.asp?id=39992 *Accessed 3 October 2014*

[23] Schaumann G, Bedacht S, Ortner A, Tebartz A, Wagner F, Spindloe C, Astbury S, Hook S, Tolley M and Roth M, Thin Cryogenic Hydrogen Targets for Laser-Driven Ion Acceleration: Fabrication & Characterisation *TFW5 conference 6-11 July 2014*

Characterisation of Diamond-Like Carbon (DLC) laser targets by Raman spectroscopy

D.Haddock[1], T Parker[1], C Spindloe[1] & M Tolley[1]

[1]Science and Technology Facilities Council, Rutherford Appleton Laboratory, Harwell Oxford, Chilton, Oxon, OX11 0QX, UK

E-mail: david.haddock@stfc.ac.uk

Abstract The search for target materials suitable for High Power Laser Experiments at ultra-low thicknesses (below ten nanometres) is ongoing. Diamond-Like Carbon is investigated as an answer for a low-Z material that can survive target chamber pump-down and laser pre-pulse. DLC was produced using Plasma-Enhanced Chemical Vapour Deposition, using with varying gas flow mixtures of argon and methane. The methane plasma deposits amorphous carbon onto the substrate and the argon plasma re-sputters the weakly bonded carbon leaving a high proportion of diamond-like bonding. Bonding natures were probed using Raman spectroscopy; analysis of the resulting spectrum showed that flow rates of 40sccm/60sccm methane to argon produced DLC films with a diamond-like (sp^3) content of ~20%. Increasing the methane gas flow decreased this value to less than 5%. DLC foils were processed into laser targets by method of float off; using a sodium chloride release layer and lowering into water, this was then lifted onto an array of apertures allowing for laser irradiation of the material with no backing. DLC with 20% sp^3 content showed superior yield when compared to other materials such as metals and some plastics of the same thickness, with ~70% of the target positions surviving the float off procedure at <10nm. As a result of this work DLC targets have been available for a number of experiments at the Central Laser Facility.

1. Introduction

Target Fabrication at the Central Laser Facility supplies solid targets to experiments at the Gemini and Vulcan Petawatt laser systems at Rutherford Appleton Laboratories. There has been a significant increase in demand for ultra-thin <10nm low Z foils for experiments in ion acceleration with potential future applications in medicine, fusion science and astronomy simulations.[1] As a response to this demand Target Fabrication gained the capability to produce Diamond-Like Carbon by Plasma Enhanced Chemical Vapour Deposition (PECVD).

Diamond-Like Carbon is an amorphous carbon film with a significant fraction of sp^3 bonding. sp^3 bonding is the covalent bonding exclusively found in diamond. "*sp*" notation donates a hybridisation of the s and p atomic orbitals with sp^3 being a single electron from the s orbital and three from the p. sp^2 bonding is two electrons in the p orbital and is found in graphite, leaving a free electron between sp^2 bonded layers. These two bonding types present in DLC are probed by Raman spectroscopy.

Sub 10nm foils inherently suffer from considerable yield issues when suspended across a window in the production phase, once put inside the chamber they are subject to pressure from the pump-down cycle of the target chamber and the presence of a laser pre-pulse. Diamond-Like Carbon is an ideal

Journal of Physics: Conference Series **713** (2016) 012007 doi:10.1088/1742-6596/713/1/012007

choice to address yield issues as it possesses the material properties of diamond but in the form of a thin film coating. [2]

DLC is produced by PECVD and the bonding nature of the resulting thin films is probed by Raman Spectroscopy. In this discussion the relationship between coating parameters and resulting bond nature is linked to the suitability for laser targets.

2. Production of Diamond-Like Carbon by Plasma Enhanced Chemical Vapour Deposition

The coating plant used was a modified RIE80 Etch system. It is fitted with a RF 600W at 13.56MHz power supply. Gas flow rate is controlled via an automatic pressure control valve at the rear of the chamber and mass flow controllers constrain the incoming gas which is introduced for the top chamber by a shower head configuration. Plasma is created by discharge between two electrodes.

For deposition of DLC; firstly the substrates are cleaned by an Oxygen plasma clean (300mbar 150W) in the PECVD plant. Next, by method of thermal evaporation, sodium chloride is coated onto the substrate. Sodium chloride acts as a release layer for the resultant DLC film. When lowered into water the NaCl dissolves and the DLC sits on the surface tension of the water, this can then be lifted onto a target array which has multiple individual target positions.

The coated substrate is placed inside the PECVD chamber for DLC coating. The plasma in process is a gas mixture of methane and argon. The methane plasma deposits amorphous carbon with some inclusion of bonded hydrogen onto the substrate, the Argon plasma re-sputters weakly bonded sp^2 carbon atoms and leaves a higher proportion of sp^3 content.

Figure 1 Modified RIE80 etch system for DLC coating. Touch screen console can be seen on the left and gas feed line can be seen on the wall at the rear. The target chamber is the black cylinder in the centre and measures 30cm diameter.

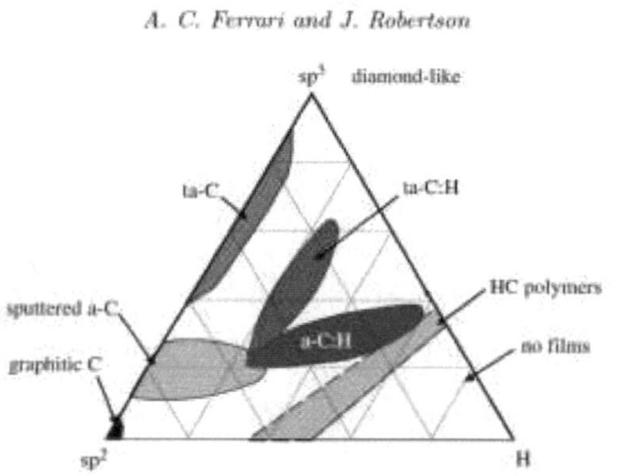

Figure 2 Ternary phase diagram of amorphous carbons. The three corners correspond to diamond, graphite and hydrocarbons respectively. Ta-C:H or tetra-hedral amorphous carbon is an alternative name for DLC. [3]

A typical coating uses a gas flow rate into the chamber of 40sccm methane and 60sccm argon, with a constant chamber pressure of 300mbar at 300W power and a process time of 300s. This produces a DLC coating less than 10nm thick.

<10nm DLC using these process parameters produces a film that is robust enough to survive open areas of 500um diameter allowing for use in laser experiments, yield across an array of 500um holes is typically 60-80% at this thickness. This is superior to any other material processed in this manner. Trailing metals such as aluminium and copper at this thickness is futile and no successful float offs are achieved. The float off method has been successful in producing foils with flatness across a 500um hole with surface deviations of only a few microns, this is achieved by using a small raised area surrounding the hole, during drying; tensile stress pulls the foil flat over the open aperture.

An alternative technique to manual float off is being developed by Christopher Spindloe at the Central Laser Facility and it is expected that this will eventually replace the manual method. This uses a substrate of silicon on which the target material is deposited, this is then etched from the rear through the silicon to leave the target foil with no backing. This has been successful with Atomic Layer Deposition deposited materials at thicknesses of ~5nm with material such as tantalum Nitride. The drawback to the method is this does not produce the flatness across the target achievable with the float off procedure; this is often a critical tolerance for laser experiments.

3. Characterisation of DLC by Raman Spectroscopy

Raman spectroscopy was used to characterise the bonding nature of the foils produced by PECVD. This technique allows the ratio of diamond-like (sp^3) to amorphous bonding (sp^2) to be estimated. Raman spectroscopy utilises the Raman shift to probe vibrational modes in a sample; the phonon vibrations are unique to the bonding present. A Raman shift is an inelastic scattering of light with vibrational modes in which the laser light is shifted (Raman shifted) to a slightly different wavelength. The background reflected light is eliminated using gratings and the Raman shift is plotted against relative intensity to give a spectrum. In visible Raman spectroscopy of diamond-like carbon the sp^3 bonding prominence cannot be probed directly as sp^2 sites are strongly resonant and therefore dominate the spectrum. However it has been shown that the relative intensities of the so called D and G peaks is indicative of sp^3 content [4].

Using a Renishaw 514nm excitation source at a power of 0.01mW the spectrum was fitted to Gaussian approximations in Origin pro 8, with which the intensities of the spectrum peaks could be compared. There are two prominent peaks one at around 1355cm^{-1} the so called D peak, which is due to the breathing of graphitic-like rings, this only occurs where the graphite rings have room to stretch, and so does not occur in continuous sheets of graphitic rings. The other prominent peak is the G peak which is around 1516cm^{-1}, this does not require the presence of 6-fold graphite rings and so occurs at all sp^2 sites. Broadly speaking I(D) (integrated intensity) increases with increasing disorder moving from graphitic dominate carbon to nano-crystalline graphite and amorphous carbon. I(G) tends to keep relatively consistent intensity but undergoes a two stage process with its relative wavenumber position. The peak position moves to lower wavenumbers with increasing bond disorder but moves higher again when creation of sp^3 chains occurs. Using these two indicators sp^3 to sp^2 ratio can be reliably inferred.[3]

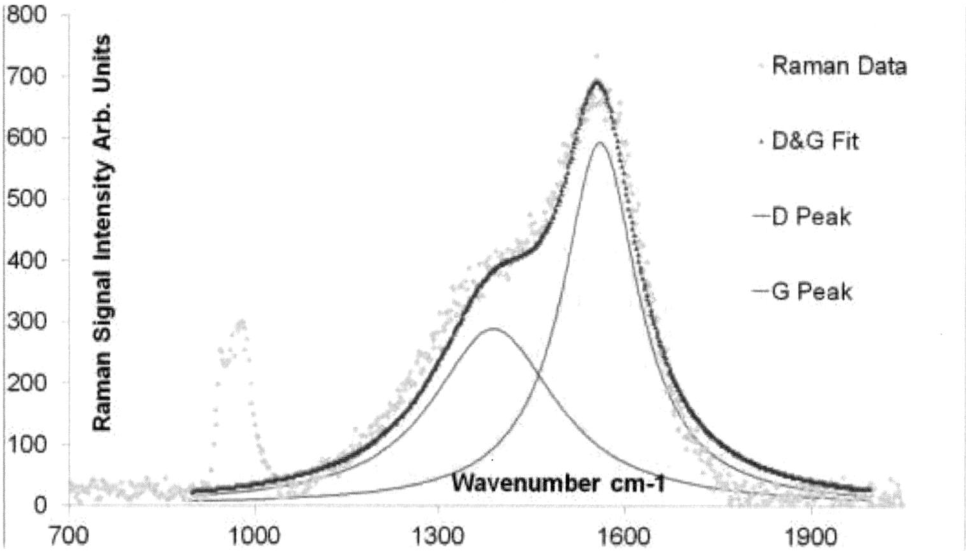

Figure 3 Typical Raman spectrum taken from the produced DLC. The smaller peak on the left around 900cm-1 is the secondary Silicon peak and is present due to the silicon substrate on which the film was coated. The D Peak is around 1355cm^{-1} and the G peak is around 1516cm^{-1}. I(D)/I(G) and the position of the G peak are the most informative factors in analysing the relative sp^3/sp^2 ratio.

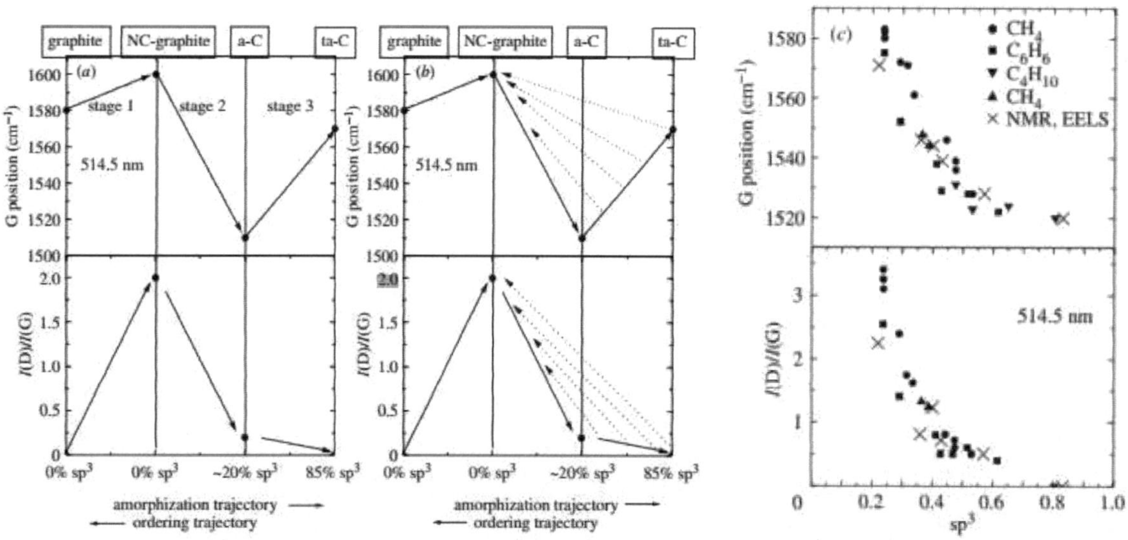

Figure 4 The three-stage model of the variation of the Raman G position and the D-to-G intensity ratio, with increasing disorder. The dotted left-pointing arrows in (b) mark the non-uniqueness region in the ordering trajectory. (c) Shows the data for as-deposited amorphous carbon collated from literature: Tamor & Vassel (1994) and Ferrari and Robertson (2000) plotted against sp^3 ratio as measured by NMR, EELS [4]

Using the data from Robertson and Ferrari 2004 in figure 4(c) a comparative analysis can be made with the DLC coatings produced by PECVD. The coating parameter of interest was the ratio of the inlet flow gases: methane and argon.

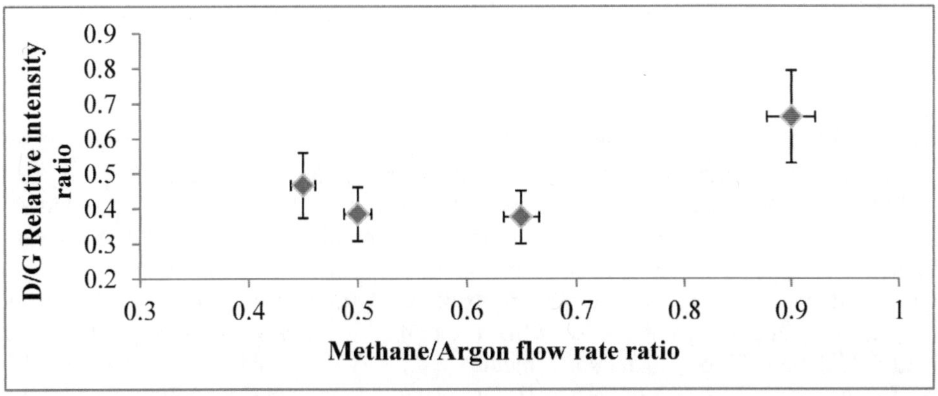

Figure 5 Diamond-Like Carbon produced with differing Methane/Argon flow rates, plotted against the relative intensities of the D and G peaks in the resulting Raman spectrum. The error in the D/G intensity ratio is high due to the error in the fitting of Gaussian peaks to the spectrum data.

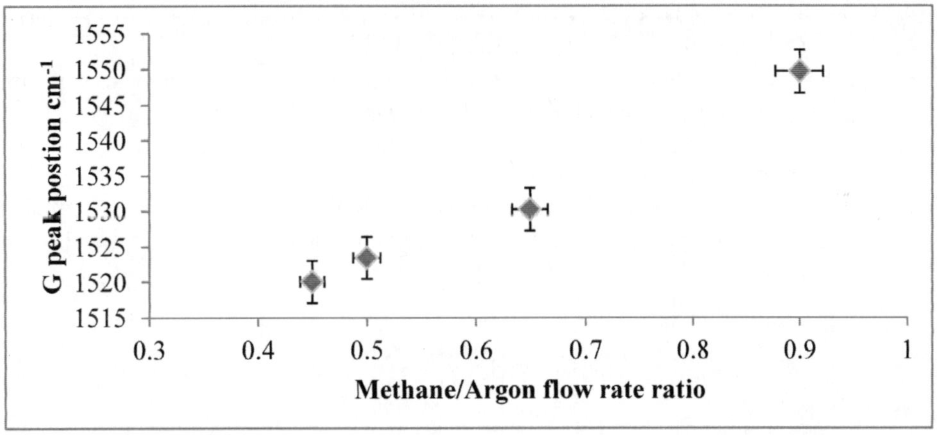

Figure 6 Diamond-like carbon produced with differing Methane/Argon flow rates, plotted against the position of the Gaussian fitted G peak in the resulting Raman spectrum.

Figures 5 and 6 confirm that decreasing the flow mixture ratio of methane to argon has a measureable effect on the resulting Raman spectrum especially for G peak position but the trend is unclear for D/G peak intensity potentially due to the errors in Gaussian fitting undertaken. Comparing the data to that shown in figure 4(c) it can be seen that the relative intensity ratio of the D and G peaks corresponds to a coating in the region of 40-60% sp^3 content. From the G peak a decrease in G peak position corresponds to an increase from 40-60% sp^3 content. This confirms predictions of a higher argon inclusion in the plasma process re-sputtering the weakly bonded sp^2 carbon and only partially ablating the sp^3 bonded carbon. [5]

The suitability for laser targets was tested by mounting foils over multiple 25 position target arrays. During the "float off" mounting procedure; it could be seen how many target positions remained complete and unbroken. It was seen that DLC coatings with a higher inferred sp^3 ratio had better yield than those of lower sp^3 ratio at ultra-thin thicknesses (~3-30nm). However, at higher thicknesses (100nm and above) the DLC coatings with high sp^3 ratio showed cracking under internal stress and had considerable yield issues. Between 30nm and 100nm no significant difference was observed in target yield. These results are purely observational over number of repeats and need to be quantified by means of strength/pressure testing.

Conclusion and future work

Diamond-Like Carbon has been successfully fabricated as a laser target for use at the Central Laser Facility and fielded on a number of experiments on the Gemini high repetition rate laser system. DLC was created by Plasma Enhanced Chemical Vapour Deposition using varying methane and argon gas flows into the process chamber. Post characterisation of the bonding present in DLC foils was carried out by Raman Spectroscopy. The analysis showed that by increasing the Argon ratio in process increased the resultant sp^3: diamond-like bonding present. By comparing bonding nature to observed survivability across a target array it was seen that the higher sp^3 bonding present produced a foil that showed better yield than that of a lower sp^3 content.

Future work aims at investigating methane in the plasma process at the 10-30% range, initial trials have seen very low coating rates at lower methane inclusion but a new power supply is being trialed.

References

[1] Kiefer D, Yeung M, Delzanis T and Foster P S. Relativistic electron mirrors from nanoscale foils for coherent frequency upshift to the extreme ultraviolet 2013 *Nat. Coms.* **4** 1763

[2] Liu Y, Erdemir A and Meletis E I. A study of the wear mechanism of diamond-like carbon films *Surface & Coatings Technology* **82** 48-56

[3] Ferrari A C and Robertson J. 2004 Raman spectroscopy of amorphous, nanostructured, diamond–like carbon, and nanodiamond *Phil. Trans. R. Soc. Lond.* **362** 1824

[4] Ferrari A C and Robertson J 1999 Interpretation of Raman spectra of disordered and amorphous carbon *American Phys. Soc.* **61** 96-106

[5] Filik J, May P W, Pearce S R J, Wild R K and Hallam K R. 2003 XPS and laser Raman analysis of hydrogenated amorphous carbon films *Diamond and Related Materials* **12** 974-978

Novel Target Fabrication Using 3D Printing Developed at University of Michigan

Sallee R Klein[1], Michael Deininger[2], Robb S Gillespie[1], Carlos A Di Stefano[1], Michael J MacDonald[1], Mario J-E Manuel[1], Rachel P Young[1], Carolyn C Kuranz[1], Paul A Keiter[1] and R Paul Drake[1]

[1]Department of Atmospheric, Oceanic and Space Sciences, University of Michigan, Ann Arbor MI
[2]Medical Innovations Center, University of Michigan, Ann Arbor MI

E-mail: salleer@umich.edu

Abstract. The University of Michigan has been fabricating targets for high-energy-density experiments for the past decade. We utilize the technique of machined acrylic bodies and mating components acting as constraints to build repeatable targets. Combining 3D printing with traditional machining, we are able to take advantage of the very best part of both aspects of manufacturing. Here we present several recent campaigns to act as showcase and introduction of our techniques and our experience with 3D printing, effecting how we utilize 3D printing in our target builds.

1. Introduction

Target fabrication is a crucial aspect of high-energy-density-physics. To enable physicist to compare results from different targets within an experimental campaign, designing and building repeatable targets is fundamental to obtaining meaningful data. Finding better ways to build targets with ever challenging geometries can be difficult. At the University of Michigan we have been designing targets comprised of acrylic outer structures that facilitate precisely built targets. Our acrylic structures enable us to build the majority of our targets by hand, using no mechanism. This technique generates targets as repeatable as our machining, reducing target to target variation substantially. Often the target design is too complex to machine as one piece. We then fabricate the acrylic outer structure in two pieces that mate in an unique configuration constraining individual components in precise orientations. In this manner, targets cannot be assembled incorrectly because the parts simply won't connect to one another. Over the years this method of building targets has been quite successful. But recently we have had to cope with the demands of shooting at smaller facilities where shot times extend over weeks, requiring targets that can number well over 100. Traditionally machining this many targets is not practical. Not only can the sheer number of targets become daunting, but designs have become ever more complex. Some of the designs required are simply not possible to machine in a traditional manner. Born from the overwhelming amount of targets needed and the increasing sophistication with target designs, we began to test how 3D printing could be of benefit to us and the target fabrication community in general.

Figure 13. This is a close-up view of our 2014 Trident X-ray Fluorescence target showing how the two cones mated together. A square stalk with square stalk hole was used to easily fix orientation of the target once placed in the target chamber. A spacer sandwiched between cones and a rod threaded through both cones constrained these components during the build.

Figure 14. Along with the spacer and rod used to constrain the cones our 2014 Trident X-ray Fluorescence target, shown close-up in Figure 13, a simple jig supported the stalk and constrained the rod that threaded through both cones and into base of the jig. It is jigs such as these that allow us to construct targets with little or no specialized equipment.

us to create designs that traditional machining could not otherwise accomplish. For the scale of targets that are used in our experiments, 3D printing alone was not often feasible. But, in navigating through the constraints that are inherent in 3D printing parts of this size and combining traditional machining techniques, 3D printing provides valuable benefits to target fabrication through its cost effectiveness and shorter manufacturing time scales.

Acknowledgments
This work is funded by the U.S. Department of Energy, through the NNSA-DS and SC-OFES Joint Program in High-Energy-Density Laboratory Plasmas, grant number DE-NA0001840, and the National Laser User Facility Program, grant number DE-NA0000850, and through the Laboratory for Laser Energetics, University of Rochester by the NNSA/OICF under Cooperative Agreement No. DE-FC52-08NA28302.

[1] Klein S, *et al*, Innovations in target fabrication techniques at the University of Michigan 2013 *FST* **63** 2 305-12

[2] Klein S R, Manuel M J-E, Pollock B B, Gillespie R S, Deininger M, Kuranz C C, Keiter P A and Drake R P Construction of a solenoid used on a magnetized plasma experiment 2014 *Rev. Sci. Instrum.* **85**

[3] MacDonald M J, *et al* Demonstration of x-ray fluorescence imaging of a high-energy-density plasma 2014 *Rev. Sci. Instrum.* **85**

[4] C. K. Li, *et al* Magneto-Inertial Fusion...Enlisting Magnetic Fields for Improved Laser-Fusion Implosions 2006 *Rev. Sci. Instrum.* **77** 10E725

X-ray tomography characterization of density gradient aerogel in laser targets

L Borisenko[1], A Orekhov[1], C Musgrave[2], W Nazarov[2], Yu Merkuliev[1], N Borisenko[1, 3]

[1] P.N. Lebedev Physical Institute of the RAS, 53 Leninskiy prospect, Moscow 119991 GSP-1, Russia
[2]University of St Andrews School of Chemistry, Purdie building, North Haugh, St Andrews KY16 9ST, UK
[3] National Research Nuclear University MEPhI, 31 Kashirskoe shosse, Moscow 115409, Russia

E-mail: borisenko.lidiya@physics.msu.ru

Abstract. The low-density solid laser target characterization studies begun with the SkyScan 1074 computer microtomograph (CMT) [1, 2] are now continued with higher resolution of SkyScan 1174. The research is particularly focused on the possibility to obtain, control and measure precisely the gradient density polymers for laser target production. Repeatability of the samples and possibility to obtain stable gradients are analysed. The measurements were performed on the mm-scale divinyl benzene (DVB) rods.

1. Introduction

Low-density micro- and nanostructured polymers – foams and aerogels correspondingly, are widely used in high-power and high-intensity laser physics. However, structures with smooth and known density profile were not yet achieved as a stable technology and would be preferable, for example, for Rayleigh-Tailor instabilities mitigation, laser energy smoothing, laboratory astrophysics studies and as a substitute for step-density targets to generate plasma profile close to the one of the corona of compresses shock-ignition target [3].

The inner structure of an aerogel consists of thin polymer fibres (or walls for foams) forming an open-cell quasiperiodic microstructure. The mean density understood as averaging the wall/fibre solid density multiplied by the solid elements volume then divided by the aerogel volume no less than several pore sizes. Experimentally the density of a solid low-density structure is mainly measured by weighting. This only provides the overall average density of the sample and gives no information on spatial changes.

Wet gel structures with density gradients were examined in [4] and the X-ray microtomography diagnostics was applied for density profile analysis of semitransparent polymer materials. Here we propose and perform precise characterization of mm-size divinyl benzene (DVB) aerogel which is close to laser target scale and requirements.

2. Samples

The low-density divinyl benzene (DVB) rods are chemically synthesized as gel structures inside capillary glass tubes then subjected to solvent exchange. Finally the samples are dried in the Polaron®

critical point dryer (CPD dryer). The height of the rods varied from 3 to 10 mm with the diameter of around 1 mm. Here chemical synthesis was fulfilled for a large bath of similar samples, so as to model gradient targets mass production. Then CMT measurements were performed to yield certain statistical data. Method of production of these density gradient foams is described in previous publication [1].

Self-standing DVB aerogel cylinders display the density gradient as visibly differing optical transparency in low- and in high-density regions. Inner microstructure and surface were examined by SEM in [1]. The material turns out to be homogeneous on the scale of several microns which increases precision of X-ray diagnostics with the voxel dimensions of around 10 um.

Besides density-gradient samples which are the main goal of our work, we needed the constant-density samples for calibration of density measurements. The procedures of chemical synthesis, solvent exchange and CPD drying for constant-density aerogels, excluding the conditions to produce gradient, remain strictly identical with gradient-density samples.

3. X-ray microtomography for density measurements

X-ray tomography is a non-destructive penetrative method for volumetric visualization and material's characterization based on the absorption properties of the material. Mathematically reconstructed flat X-ray images show absorption in each spatial point of the sample:

$$I_{image} = I_{initial} \cdot e^{-\mu_{abs} \cdot x},$$

where I_{image} intensity is read from CCD as grayscale index, μ_{abs} – absorption coefficient for the specific material, x – optical path through the sample.

3.1. SkyScan 1174

The microtomograph SkyScan 1174 used in this work allows the maximum resolution of 6.9 um per pixel. However for DVB rods scanning a 17.5 um resolution was chosen to fit the view window. Tungsten cathode of the X-ray tube provides broad-spectrum X-ray radiation in cone beam geometry. The sample is placed onto a rotating table B (Figure 1a) on the way of X-rays to form multiple flat absorption images (Figure 1b) by CCD camera. The rotation step could be as small as 0.6 degree. The set of images is reconstructed to 3D model using special software by Feldkamp [5] algorithm. The reconstruction gives a number of horizontal cross sections with absorption recalculated to internal computer tomography units (CT number) in each voxel of each cross section. The obtained CT number in each voxel corresponds to grayscale index of the 2D image and thus through X-ray absorption in voxel to the local density.

Figure 1a. SkyScan 1174 inner view. A – X-ray tube window; B – rotating table with a DVB sample; C – *Be* filter of the CCD camera window.

Figure 1b. Flat X-ray image of DVB rod.

Journal of Physics: Conference Series **713** (2016) 012009 doi:10.1088/1742-6596/713/1/012009

3.2. Absolute calibration

The reconstruction with the built-in SkyScan 1174 software does not give information on absolute density of the sample. However one can only analyse comparatively the exposed density-gradients with the obtained statistics on CT numbers. This is not enough for purposes of laser target characterization. Therefore the absolute calibration of the CMT measurements was performed using the samples with initially known average densities. The average densities were calculated from the solution concentration and checked by weighting.

For the calibration samples of pre-known density separate CMT scanning was done. Averaging over each horizontal cross section we got a distribution of CT numbers over the height of each sample (Figure 2). CMT-measured density variation via height for cylindrical samples of 4 constant nominal densities are plotted in solid line, average CT numbers of constant-density derived from real curves are plotted in dash lines.

Averaging was carried out by the following method: the cross section was scanned for non-zero pixels and the remaining objects were eroded by several pixels to minimize the boundary artefacts. Then, the resulting group of pixels was used to calculate the average CT number.

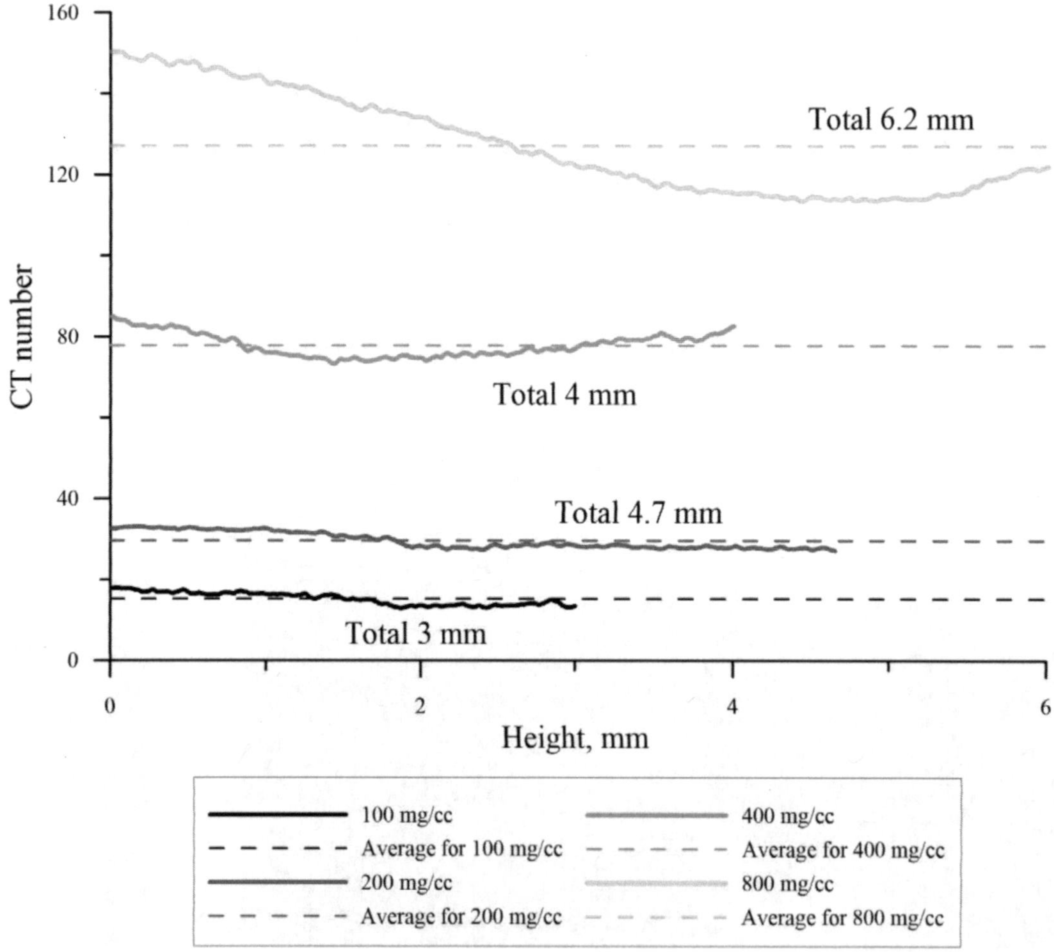

Figure 2. Density variations (in units of CT numbers) over height of the sample for the constant nominal density DVB rods. Dashed constants are calculated average values.

Absorbing properties of materials in X-rays strongly depend on the nuclei charge Z. Thus for each atomic composition the calibration is to be performed independently.

The calibration curve for DVB aerogel is shown in Figure 3 and defines the conversion coefficient from CT numbers to density of 0.16 ±0.01 units/(mg/cc). The aerogels with density gradient are further analysed in terms of reconstructed and recalculated density.

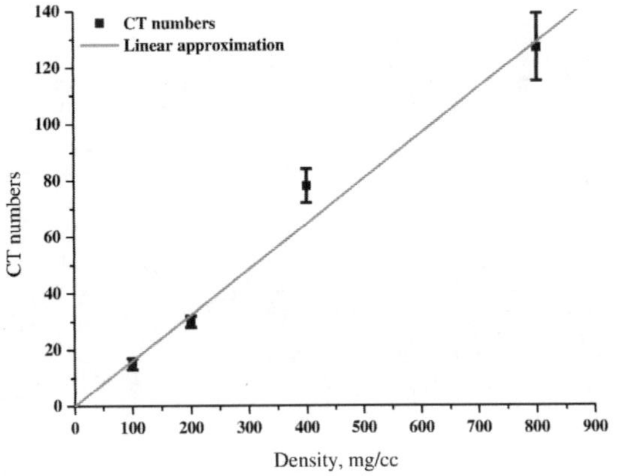

Figure 3. CT numbers to density calibration curve.

4. Density gradient in DVB aerogel rods: realisation and measurements

The density gradient DVB rods were examined on SkyScan 1174 with the same scanning parameters and resolution as the calibration samples. The data was reconstructed with the built-in software, averaged CT numbers were calculated over each cross section and then the obtained distribution was recalculated in terms of density in mg/cc.

To diminish the noise and for stricter curve analysis the density function was plotted after application of moving average filter over 11 points (Figure 4a left, Figure 4b left). To visualize the density changes over the height 1st derivative was calculated and is further referred to as gradient function or gradient. The gradient was plotted and analyzed with Fast Fourier Transform (FFT) filter over 15 points to cut off the high-frequency noise.

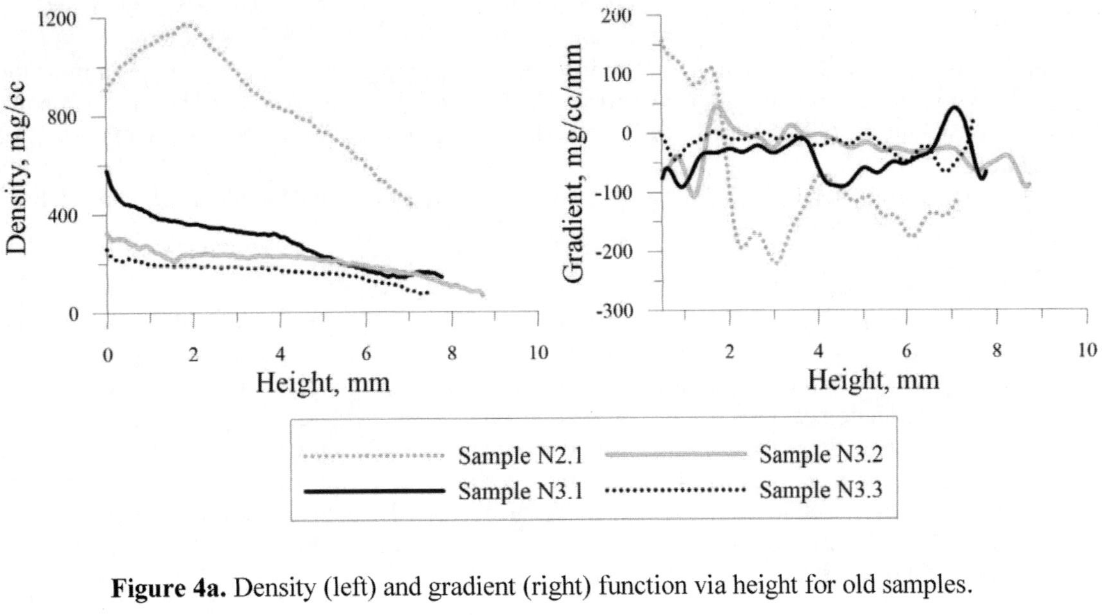

Figure 4a. Density (left) and gradient (right) function via height for old samples.

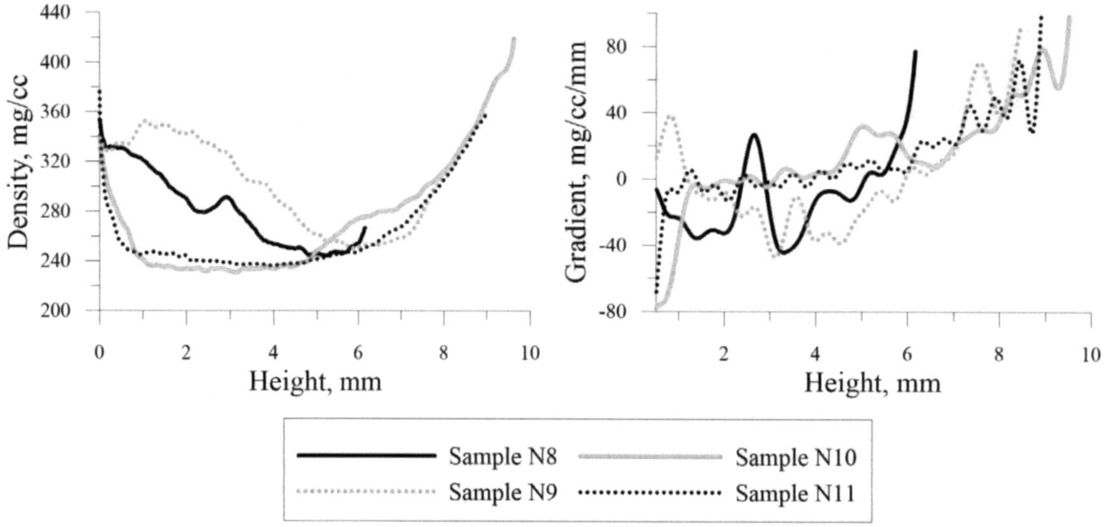

Figure 4b. Density (left) and gradient (right) function via over height for new samples.

We used 2 batches of samples with almost two year gap in the time of fabrication. The Figure 4a refers to the earlier done samples, the same as described in [1]. After 2 years of storage they were measured once more, now together with newly fabricated samples of Figure 4b for comparison, production studies and more profound characterization with next-generation CT-scanner SkyScan 1174.

Comparing the old (Figure 4a) and the new (Figure 4b) samples we found no significant difference. The samples are stable in time and can be stored and transported.

For previous results [1] we used practically manual calculations (averaging and part of filtering procedures). Thus the precision was less than possible but the route and algorithms for future standard measurements were justified. For example, the observed waves of density to some extend were due to calculation inconsistences. Larger pool of experimental data processed at present with higher accuracy for every gradient sample, provides reliable data and this is still comparable with previously measured results.

On Figure 4 we observe the regions of almost constant average gradient of 10-20 mg/cc/mm on the samples N3.1 (from 2 to 5 mm from the bottom), N3.2 (from 3.5 to 7 mm), N3.3 (from 1.5 to 4 mm), N9 (from 6 to 7 mm) and N11 (from 4 to 7 mm). Hence such gradient is reproducible on the scale of up to 3 mm.

The maximum gradients obtained (sample N2.1 on Figure 4b) are 150-200 mg/cc/mm and are realized within 1.5 mm length.

5. Results and discussion

The microtomography technique was proposed and used in this work for study and characterization of density gradient plastic aerogels. Self-standing mm-scale aerogel rods are a good approach for further analysis of laser targets. Standard procedures are worked out for profound studying of separate samples during technological research and for routine monitoring of cylinder-shape samples.

Obtained and precisely analyzed density and density gradient distributions over height of the DVB rods showed stable gradient regions of up to 4 mm. Maximum gradients of 150-200 mg/cc/mm were observed on 1.5 mm length interval. The average gradient was calculated 20 mg/cc/mm. The gradients are reproducible on the scale of 3 mm although no dependence of the gradient sharpness on the length of the sample was observed.

The DVB samples with density gradient are stable in time (years) and the technique of density gradient aerogel production is promising for laser target development. More importantly, it has been demonstrated that the production method for density gradient foams is reproducible and precise positioning of necessary gradient interval for further isolation is possible and easy with CMT.

References

[1] Borisenko N G, Nazarov W, Musgrave C S, Merkuliev Y A, Orekhov A S and Borisenko L A (2014). Characterization of divinyl benzene aerogels with density gradient using X-ray tomography technique. *Journal of Radioanalytical and Nuclear Chemistry*, *299*(2), 961-964.

[2] Akimova I V, Akunets A A, Borisenko L A, Gromov A I, Merkuliev Y A and Orekhov A S (2014). Metals produced as nano-snow layers for converters of laser light into X-ray for indirect targets and as intensive EUV sources. *Journal of Radioanalytical and Nuclear Chemistry*, *299*(2), 955-960.

[3] Goyon C, Depierreux S, Yahia V, Loisel G, Baccou C, Courvoisier C, and Labaune C (2013). Experimental Approach to Interaction Physics Challenges of the Shock Ignition Scheme Using Short Pulse Lasers. *Physical review letters*,*111*(23), 235006.

[4] Borisenko N G, Akunets A A, Artyukov I A, Gorodnichev K E and Merkuliev Y A (2009). X-ray tomography of growing silica gel with a density gradient. *Fusion science and technology*, *55*(4), 477-483.

[5] Feldkamp L A, Davis L C and Kress J W (1984). Practical cone-beam algorithm. *JOSA A*, *1*(6), 612-619.

Journal of Physics: Conference Series **713** (2016) 012010 doi:10.1088/1742-6596/713/1/012010

Automated Production of High Rep Rate Foam Targets

F Hall[1], C Spindloe[1,2], D Haddock[2], M Tolley[1,2] and W Nazarov[3]

[1]Scitech Precision Ltd, Rutherford Appleton Laboratory, Harwell Oxford, Oxfordshire, OX11 0QX, UK
[2]Science and Technology Facilities Council, Rutherford Appleton Laboratory, Harwell Oxford, Oxfordshire, OX11 0QX, UK
[3]School of Chemistry, St Andrews University, St Answers, Fife, KY16 9ST, UK

E-mail:chris.spindloe@scitechprecision.com

Abstract.

Manufacturing low density targets in the numbers needed for high rep rate experiments is highly challenging. This report summarises advances from manual production to semi-automated and the improvements that follow both in terms of production time and target uniformity. The production process is described and shown to be improved by the integration of an xyz robot with dispensing capabilities. Results are obtained from manual and semi-automated production runs and compared. The variance in the foam thickness is reduced significantly which should decrease experimental variation due to target parameters and could allow for whole batches to be characterised by the measurement of a few samples. The work applies to both foil backed and free standing foam targets.

1. Introduction

With the recent developments in laser technology there are an increasing number of high repetition rate systems that are available for utilisation by the plasma physics community for experiments investigating topics such as ion acceleration. One of the most interesting types of targets in this field is a low density target that can be used to investigate hole boring[1]. One such type is low density polymeric foam that is on the order of 10-100 microns in thickness. It is however difficult to manufacture thin foams in the numbers that are required to carry out these experiments. Current manufacturing techniques rely on a manual fill of target geometries under a microscope with a syringe and subsequent processing using a critical point dryer. We report on a production method that is using a semi-automated dispensing and curing system and developments to maximise the output of the drying process to produce targets that are technically challenging and previously not possible to fabricate.

Typical target geometries that have been requested to date have been foam densities in the range 3 to 300mg/cc and thicknesses in the range from 50 to 500 microns. For low repetition rate experiments on facilities such as the Vulcan laser at the Rutherford Appleton Laboratory (RAL), there may be a need for 10's of targets but for medium repetition rate experiments, such as those on the Astra Gemini laser at RAL there may be the need for several 100's of targets. There is a trend within the community to require the more demanding thin and lowest density targets and these specifications are especially challenging to achieve for free-standing foams and may well be beyond the realms of existing technology and beyond the budgets of experimental campaigns.

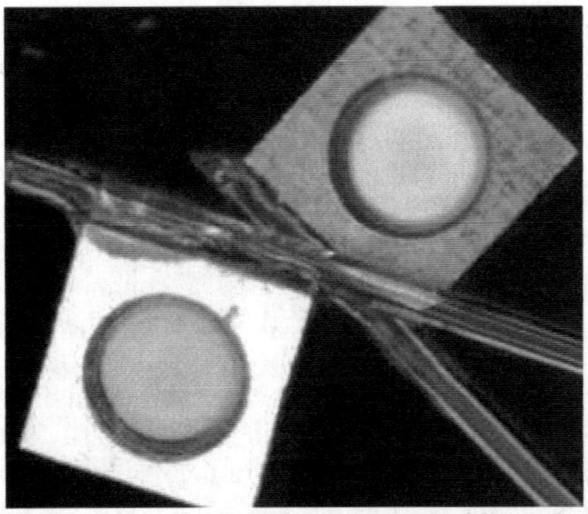

Figure 1a. Individual foam targets with washer supports for low rep rate experiments. Shot on the Vulcan laser at RAL, UK (May 2013).

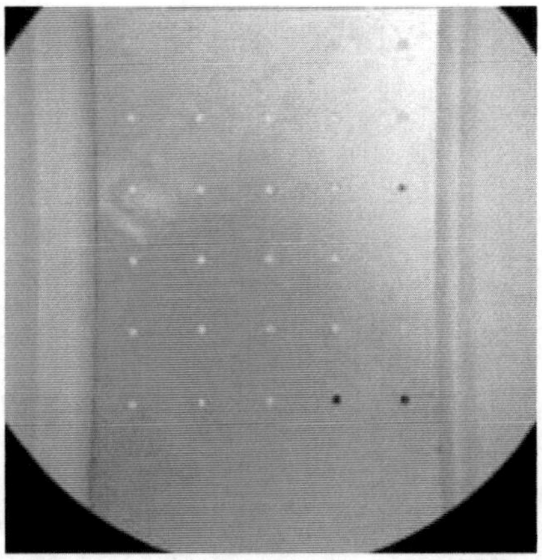

Figure 1b. Array based foam targets for med rep rate experiments. Shot on the Astra Gemini Laser (May 2013).

The composition of different foam types makes them more or less suitable for different experiments. This can be quantified in terms of the chemical makeup, the physical structure (with pore size being a key factor), density regime, machinability and finally quality. This last attribute, quality, can include uniformity, defects both in terms of high density inclusions or holes, and the presence of a higher density skin.

TMPTA or Trimethylolpropane triacrylate has a chemical composition of Carbon, Hydrogen and Oxygen. For some experiments the presence of oxygen prevents this type of foam from being used. However, the foam benefits from having no predefined pore size meaning that the density can be reduced without introducing complicating large voids. If pure chemicals are used and clean conditions are employed then inclusions of high density defects and skin effects can be minimised. The foam is not particularly easy to machine so a support or mount is used to define the shape and size of the foam.

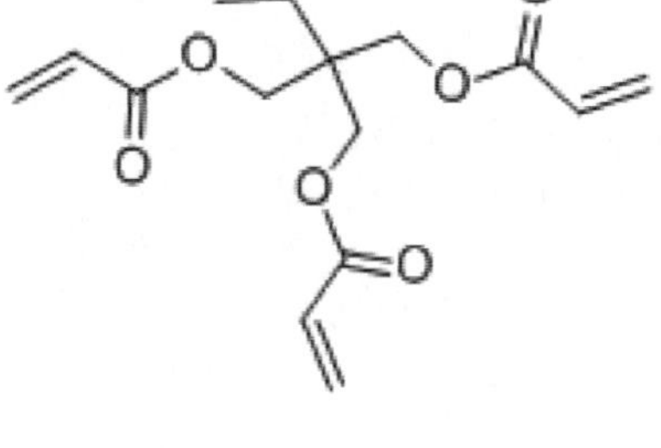

Figure 2. The chemical structure of Trimethylolpropane triacrylate (TMPTA) showing the three functional acrylic groups available for polymerisation.

2. Manual production

Foam production has an established process [2] that starts by producing a foam solution. The solution is made up of three parts. The ratio of TMPTA and a solvent defines the final density of the foam. A very small quantity of UV sensitive photo initiator is added which allows the foam to polymerise. The solvent and photo initiator are carefully chosen to only contain the same elements as the TMPTA to avoid any chance of cross contamination. Some initiators as an example might contain high Z elements such as chrome (Cr) or Iron (Fe). The foam solution is dispensed into cavities which could be individual mounts for single shot experiments or arrays for medium rep rate. This is typically done under a microscope due to the small size of the cavities and this would typically take 10 seconds per cavity. For several hundred targets this stage could take a number of hours.

A mercury arc lamp or LED UV source is used to cure the foam. Over a time frame of a few seconds, depending on the power of the UV source, the clear solution will turn white. This is because the polymer network now scatters the light that passes through and this tells us that the curing stage is complete. Again this will take several hours for several hundred targets. The last stage in the process involves removing the solvent. Drying of the foam by evaporation of the solvent would collapse the delicate foam structure. For this reason a Critical Point Drier (CPD) is used. It is a commonly used tool within industry for drying delicate samples and includes putting the targets into a high pressure vessel. Liquid carbon dioxide is used to fill the chamber at a high pressure (above 75 Bar) and the temperature is controlled from around 20°C up to 40°C. At the higher temperature the carbon dioxide behaves as a supercritical fluid displaying the properties of both a gas and liquid simultaneously. Once in this state, the carbon dioxide can be vented off without any sudden change in density and without any surface tension effects. When this process is complete, the foam is dry and ready to be used as a target. The last stage takes approximately two hours.

2.1. Polymerisation process

When UV light of an appropriate wavelength is incident on the dissolved photo initiator molecules they split into two or more free radical parts. The free radicals can either recombine but will split again under the UV or they can react with the TMPTA monomer molecules but not the carefully chosen solvent. TMPTA has three functional groups containing a carbon double bond which are possible sites for a free radical to bond with. When a free radical reacts with a TMPTA molecule it opens up the double bond and forms a small chain at this point but leaves the other half of the double bond as a free radical. It is this property that allows the chain to react with further TMPTA molecules creating longer and longer chains. Since TMPTA has three functional groups it is possible to have cross-linking between adjacent chains and this gives the overall polymer network rigidity.

High purity chemicals are key to the success of the polymerisation process because the free radical photo initiator parts or polymer chains will react with any impurity stopping the formation or growth of the polymer network. Having clean cavities is also key to the foam network bonding with the walls and therefore remaining in place once the drying process is complete.

3. Semi-Automated Production

The most time consuming stages of the foam production process are the filling and curing steps. Given the regular spacing and grid-like nature of arrays, it was thought that a robot might be able to complete both of these steps simultaneously allowing fabricators to be freed up for other tasks. The shot-to-shot variation in the foams should also be reduced by using pre-set dispensing values.

A three axis (x, y, z) robot with dispensing capability (figure 3) was used to prove the principle of automated production. The robot can be controlled via a laptop and custom written software which would allow the integration of the UV curing stage using a digital signal to trigger a LED source.

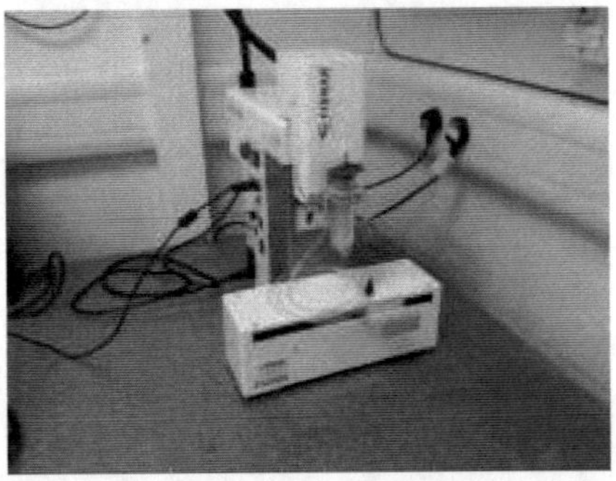

Figure 3. Dispensing robot with three axis stage. Four mounting screws are available for custom made plates.

There were a number of challenges that needed to be addressed to commission the system to be suitable to make foam targets. An array mount holder, shown in figure 4, was designed to hold multiple arrays with an x, y positional accuracy of better than 200um. When thin (<150um thick) arrays were placed into this holder, it was found that they could bow with a change in height of several hundred microns. In addition to this, different thickness array mounts could in practice be used and so it was decided that it was necessary to incorporate a feedback loop between the metal dispensing tip and the metal arrays. This would trigger only when the tip touched the array and this position in z would be considered zero. From here, the tip could then rise by a predefined amount before performing a dispense routine. This feedback loop allows the system to cope with multiple target designs and builds flexibility into the production process.

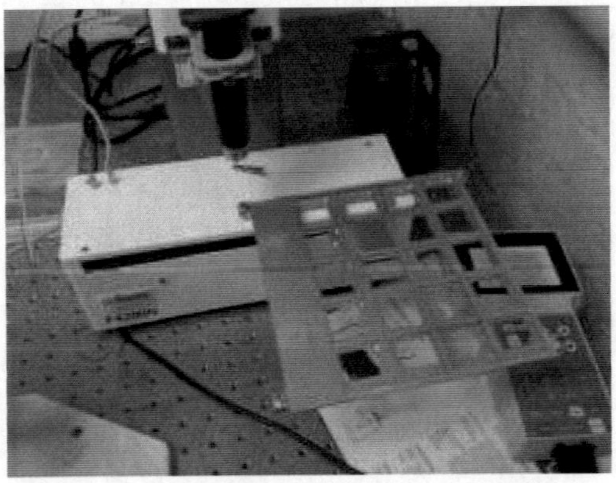

Figure 4. Holder plate with space for 20 array mounts. Positional accuracy was specified to be better than 200um in x & y.

It was discovered that when dispensing different liquids through a standard tip the physical characteristics of the liquid can yield surprisingly different results as shown in figure 5. For water, the droplet would form below the end of the tip. For the foam solution, the liquid would climb up the outside of the tip. When dispensing into a shallow container there are obvious limitations if the liquid does not present itself at the end of the needle, especially if the foam solution is to be dispensed onto a very thin backing foil of a few 10's of nanometers thick, where any contact between the needle and the foil will destroy it. It was decided to bend the end of the tip by 90° and use the wicking property of the foam solution to help with the filling procedure. With the foam solution exposed proud of the needle it could be dispensed on one side of an array hole and the meniscus dragged over to the opposite side without the needle touching the fragile foils. To ensure this is the case a gap is required between the tip and the array of the order of 50 – 100 microns which was possible to program using the contact feedback loop.

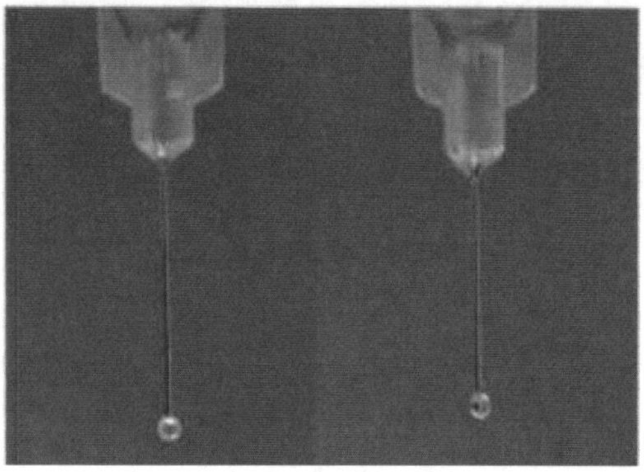

Figure 5. Left: A water droplet hangs below the end of the dispensing tip. Right: A foam solution droplet climbs up the dispensing tip.

4. Results

The height of the foam was measured relative to the flat surface of the array mount and this was done for a previously produced 36 shot array of foams made by hand and compared with a set produced using the robot. Measurements were taken using a white light interferometer which produced a 2D height map as shown in figure 6. This height map was levelled and a lineout taken through the centre of a foam to obtain a measurement.

The robot dispensing system has the ability to vary the amount of foam solution that it dispenses and therefore can vary the height of the foams above the mount that it produces, however for this investigation it was kept a constant to investigate variation over the range of positions using a common setting.

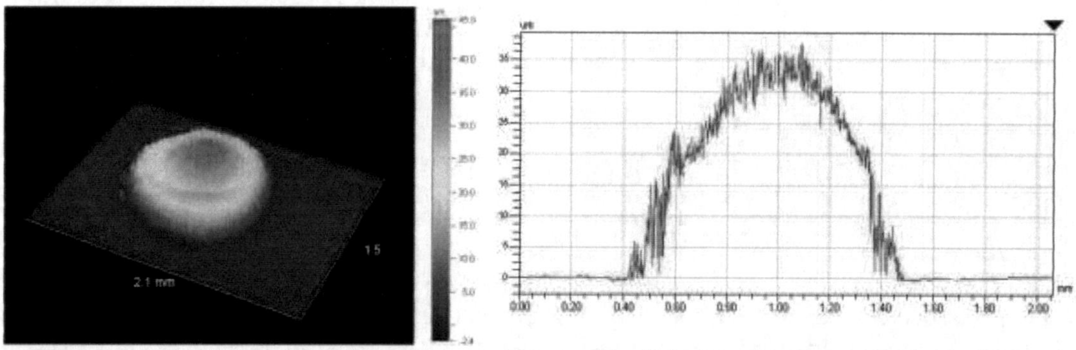

Figure 6. Left: 3D reconstruction of a scan using a white light interferometer of a foam. Right: A lineout taken through the centre of the foam to give a height profile above the level of the array mount.

The results in figure 7 show a standard deviation of 10.8um from the foams produced by hand and 6.7um from the machine made foams. Qualitatively the shape of the machine made foams (figure 8) is slightly egg shaped but consistent across the array. The shape could be modified by using a different fill routine.

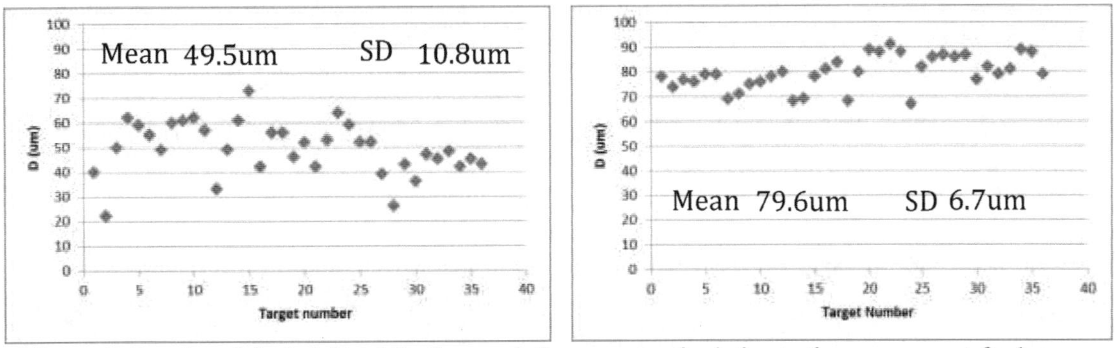

Figure 7. Results of the height of the foam solution measured relative to the array mount for human fill (left) and machine fill (right).

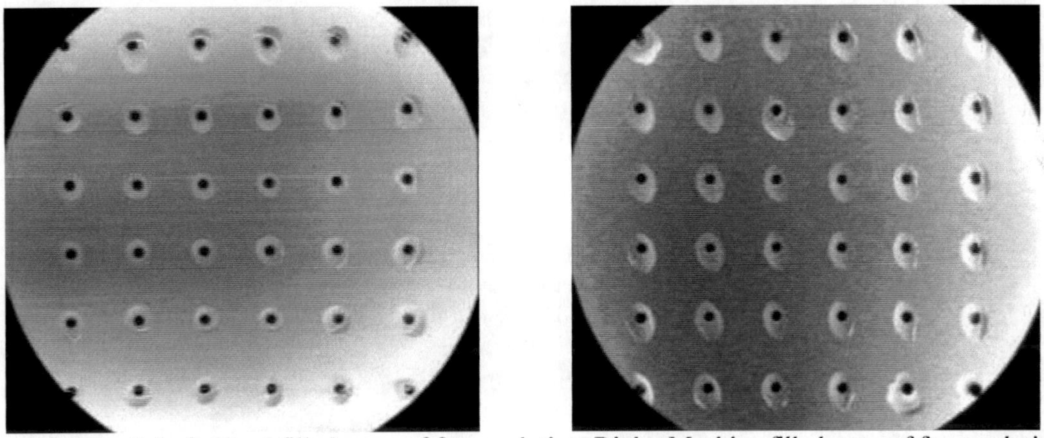

Figure 8. Left: Hand filled array of foam solution. Right: Machine filled array of foam solution.

5. Conclusions

The robot has shown good results in improving the quality of the foams by reducing the target-to-target variation and with careful programming it is hoped that this variation can be further reduced. The aim is to obtain a variance low enough that a whole batch of foams can be characterised by the measurement of only a few samples. The time taken to produce the foams has been significantly reduced. As more laser systems utilise the high repetition capabilities that are being developed a large number of arrays will be required. This time saving will be essential in ensuring that the laser systems are not target limited. Over the next few years we will develop further target geometries in collaboration with user groups in an effort to enhance the experiments undertaken.

References

[1] Robinson A P L et al, *PPCF*, **51**, 024004, (2009)
[2] Nazarov W, *Fusion Sci. & Technology* **41**, 193, (2002)

Journal of Physics: Conference Series **713** (2016) 012011 doi:10.1088/1742-6596/713/1/012011

A novel microfluidic system for the mass production of Inertial Fusion Energy shells

N T Inoue

Cardiff University, School of Engineering, Queen's Buildings, The Parade, Cardiff CF24 3AA, UK

E-mail: tatsunosukenti@gmail.com

Abstract. A system which can mass produce millimetre sized spherical polymer shells economically and with high precision will be a great step towards the Inertial Fusion Energy goal. Microfluidics has shown itself to be a disruptive technology, where a rapid and continuous production of compound emulsions can be processed into such shells. Planar emulsion generators co-flow-focus in one step (COFON) and cascaded co-flow-focus (COFUS) enable millimetre compound emulsions to be produced using a one or two step formation process respectively. The co-flow-focus geometry uses symmetric and curved carrier fluid entrance walls to create a focusing orifice-minima and a carrier flow which aids movement and shaping of the dispersed fluid(s) towards the outlet, whilst maintaining operation in the dripping regime. Precision concentric alignment of these compound emulsions remains one of the greatest challenges. However steps to solve this passively using curved channel modulation to perturbate the emulsion have shown that rapid alignment can be achieved. Issues with satellite droplet formation, repeatability of the emulsion generation and cost are also addressed.

1 Introduction

Inertial Fusion Energy IFE has shown promise since KMS Fusion's endeavours in the early 1970's[1]. In order to get a commercially viable system, challenging technical hurdles need to be overcome such as the development of high-power high-repetition driving schemes e.g. LASERs, target injection with tracking and beam-steering systems[2] among many others. Another key issue is the economical mass production of high quality spherical polymer shells. The study discussed herein describes improvements in the latter area where a continuous production of emulsions are processed into such shells using microfluidics. Current methods use a batch process for formation and polymerisation of shells which then undergo extensive sorting and lengthy characterisation to find a shell which meets the exacting specification[3]. Because of the stringent metrology, the associated costs are extensive so an extremely reliable production method, where individual characterisation is unnecessary, is the ultimate goal. One can exploit the characteristics of microfluidic flow, for example the low Reynolds numbers that are commonly observed occurring far below the turbulent flow threshold of ($Re \approx 2000$) enables laminar flow (Figure 1). Furthermore a high surface to volume ratio allows rapid chemical reactions to be performed.

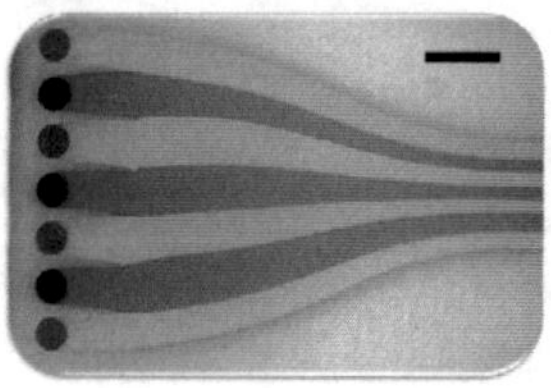

Figure 1: Laminar flow within a microfluidic channel ($Re \approx 0.05$). Alternating streams of mineral oil dyed blue, intersecting within a machined channel of polytetrafluoroethylene PTFE and with a film lid of fluorinated ethylene propylene FEP. Channel depth of 1mm and scale bar is 2mm.

When two or more immiscible fluids at certain flow rates are delivered into a specific geometrical configuration, segmented flow conditions with serial trains of the fluids result. These are called emulsions where one fluid is dispersed within another, known as a carrier fluid. Three common geometries or 'junctions' can be employed to create such emulsions (Figure 2).

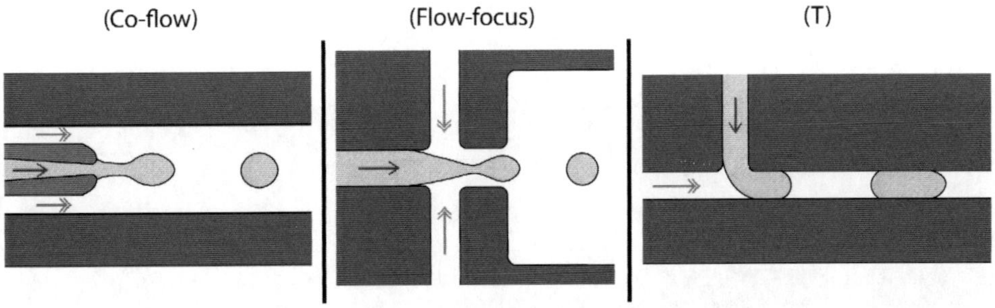

Figure 2: Three common microfluidic emulsion-forming junctions showing the dispersed fluid flow (single-point arrow) broken into droplets under the influence of the carrier fluid flow (double-point arrow).

These junctions can be operated in one of two flow regimes, called dripping and jetting (Figure 3). It has been shown however that the jetting regime is more sensitive to external factors such as vibration and changes in temperature[4][5]. A tighter distribution of emulsion volumes, expressed as the monodispersity of diameters, is found when junctions are operating in the dripping regime[6][7].

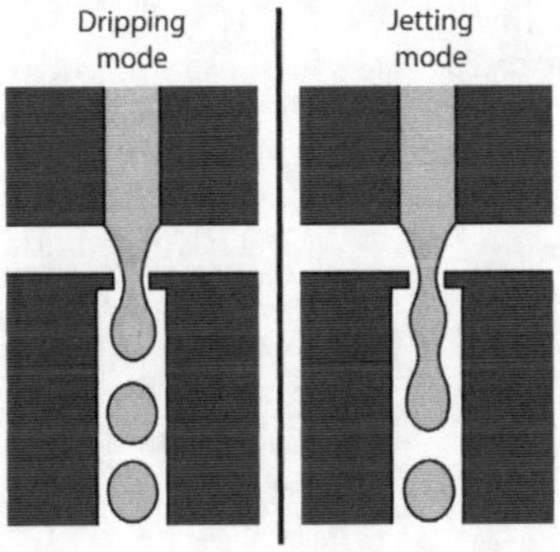

Figure 3: Emulsions being formed in a flow-focus type junction with two different regimes of dispersed fluid flow.

2 Compound emulsion generation

Compound emulsions with their multiple encapsulated regions can be formed if two or more junctions are cascaded. For current IFE research the spherical shells are produced from a double compound emulsion which consists of a central inner fluid surrounded by an outer fluid which, in turn, is suspended in a carrier fluid (Figure 4). Where a liquid monomer is used as the outer fluid, this can be polymerised to form a hard and liquid-filled shell.

When two junctions are cascaded, the device can operate using either a one or two-step droplet formation process. The attention is at the first junction, which can either be driven to form segmented flow (Figure 5 (A)) or to form co-flow (Figure 5 (B)). The dispersed fluids can then be broken off to form compound emulsions at the second junction.

Great care must be taken to ensure the junction's channel walls have the correct fluid to wall adhesion for the intended emulsion. For example, if a water dispersed in oil (W/O) emulsion is desired, the oil must

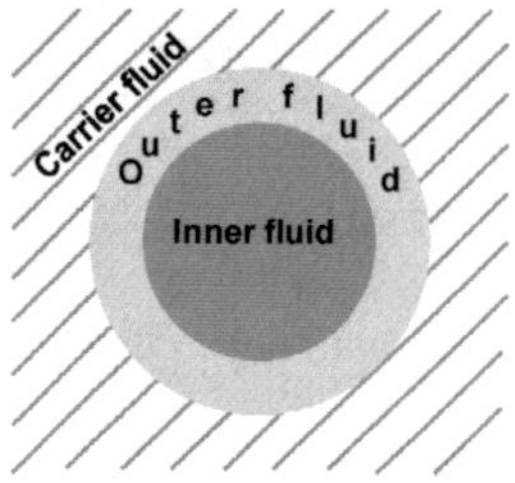

Figure 4: Double compound emulsion structure.

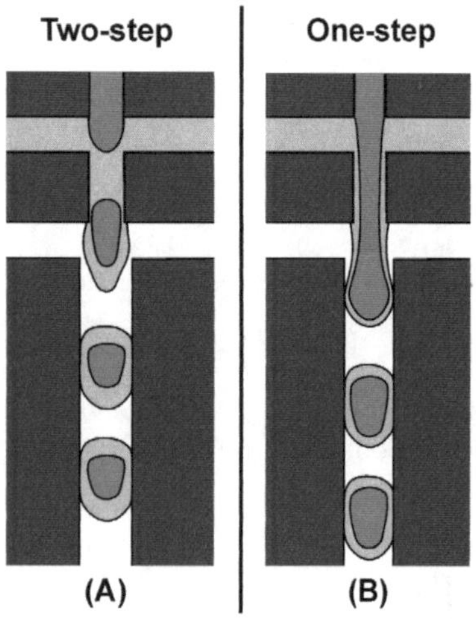

Figure 5: Illustrating a cascaded co-flow microfluidic junction forming compound emulsions using a two(A) or one(B) step process. Attention is at the first junction which can be driven to produce segmented flow or co-flow before entering the second.

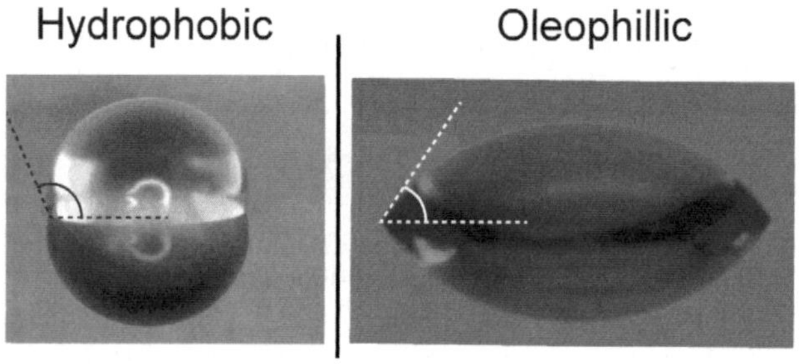

Figure 6: A droplet of de-ionised water (left) and mineral oil (right) is placed on a polished surface of PTFE in standard laboratory conditions. The Sessile droplet formed shows the material's hydrophobic repulsion and oleophillic attraction, resulting in a big difference of contact angle which enables stable production of water in oil emulsions if used as the junction's channel wall material.

preferentially adhere to the junction's channel walls rather than the water. Figure 6 shows the differing contact angles that result when either de-ionised water or mineral oil is placed on a polished surface of PTFE. The material is hydrophobic due to its low adhesion to water and therefore a high contact angle results but is also oleophillic with its adhesive affinity to mineral oil and hence has a low contact angle. It is the large difference in contact angle between the fluids towards the channel material that makes it ideal for the stable production of O/W emulsions. For compound emulsions where an oily monomer is used as the outer fluid, the configuration can be either water in oil in oil (W/O/O) or water in oil in water (W/O/W).

It is important to understand that millimetre size emulsions are much larger than is commonly associated with microfluidics. Because of the larger mass, inertial forces start to have a greater influence over that of interfacial tension, which dominates in the micrometre domain, thereby increasing the difficulty in their production. It would therefore be appropriate to say we are using millifluidics.

IFE shell manufacture using emulsions was investigated around 1979 by U. Kubo and others. However the flask-based agitation method used to make the emulsions was unsuitable for the required repeatability[8]. The process was greatly improved by M. Takagi with his invention of the triple orifice generator[9] which uses a co-flow type geometry and can produce repeatable high quality shells as shown by K. Nagai[10]. The geometric variability built into the design allows for a large range of emulsion diameters of around

0.5 - 6mm although the geometry can be simplified to a static planar design with the consequence of reducing the range. A recent study which used an enlarged cascaded T-junction[11] was shown to produce emulsions of 1.4mm to 2.5mm with a wall thickness between 0.5-0.75mm respectively albeit not the thin-walled shells (0.1-0.25mm) required for IFE. The flow into a T-junction is one-sided resulting in droplets with asymmetric internal flow eddies which could ultimately affect the shells quality. When choosing a symmetric flow junction however, extreme care must be taken to ensure that the flow and fluid entering each side is similar in all characteristics.

2.1 Setup of the microfluidic system

Microfluidic geometries are machined into plastic discs called 'chips'. A transparent lid and quartz glass window is then placed on top and the whole assembly is compressed together using a bolted stainless steel captive housing with pipes transferring fluid from beneath (Figure 7). If the disk is chosen to be polytetrafluoroethylene PTFE with a fluorinated ethylene propylene FEP film lid, the hydrophobic material properties allow for W/O/O type emulsions to be formed without the need for surface modification or surfactant. The fluids used comprise of an inner water solution, density-matched to that of the outer fluid - trimethylolpropane triacrylate (TMPTA) monomer with an added photo-initiator as suggested by W. Nazarov[12]. Mineral oil acts as the carrier fluid. Once the emulsions are formed at the junction(s) they can be polymerised using ultraviolet (UV) light to form spherical, liquid-filled shells. Photo-polymerisation occurs in a photo-reactor which can be situated either on the chip or separately.

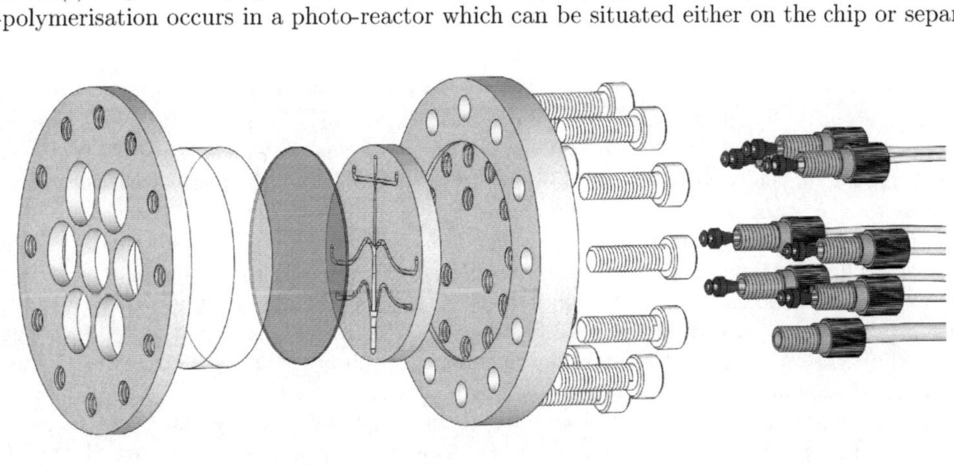

Figure 7: An exploded-view of the microfluidic chip within a stainless steel housing. A transparent lid and quartz glass viewing window are placed on top of the chip with fluid transfer pipes connecting from beneath.

2.2 The cascaded Co-Flow-Focus (COFUS) junction

The cascaded COFUS junction uses a two-step droplet breakoff process created when two separate junctions are cascaded together (Figure 8). Both are co-flow-focus COFUS geometries which are a hybrid between the co-flow and flow-focus type junctions (Figure 2). The curves redirect the carrier fluid flow from being perpendicular to the dispersed fluid into a shallow angle, like that of the co-flow. The lower curve allows a smooth translation of carrier fluid through an arc, to which the two sides form a flow-focusing orifice minima. The geometric shape of the COFUS junction therefore allows the dispersed fluid(s) to extend far enough into the junction to form millimetre sized emulsions while maintaining operation in the dripping regime (Figure 9). **(COFUS Video)** shows the junctions producing compound emulsions at a rate of 2.5Hz.

Journal of Physics: Conference Series **713** (2016) 012011 doi:10.1088/1742-6596/713/1/012011

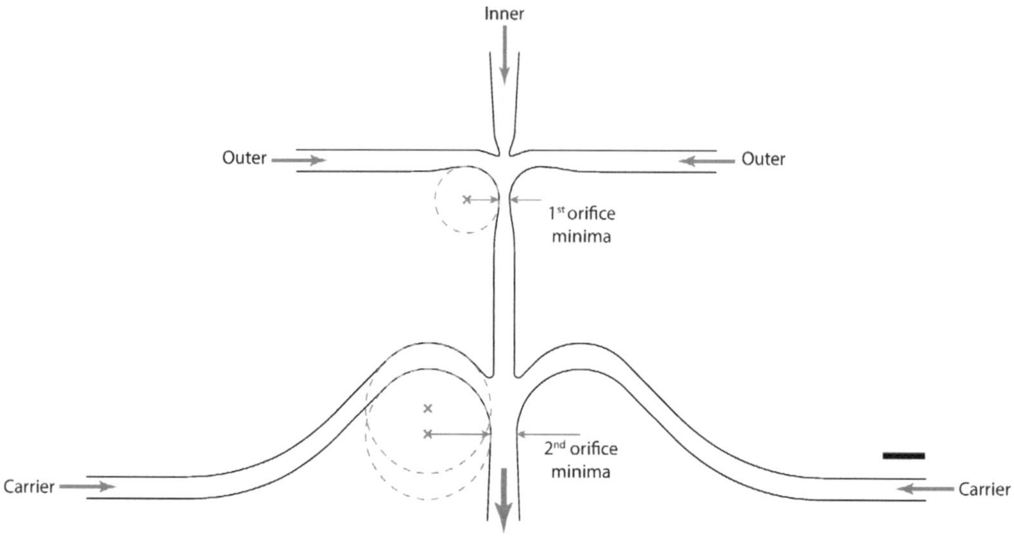

Figure 8: The cascaded co-flow-focus (COFUS) planar compound emulsion generator. Depth is to be 1mm. Scale bar is 2mm.

It should be noted that millimetre compound emulsions can also be formed using an enlarged and cascaded flow-focus geometry (Figure 10); but under some flow conditions it can start to perform undesirably by breaking up the compound emulsion and forming satellites. Compared with the average flow velocity of fluid in the carrier channel, a high velocity flow is observed around the sharp, lower corners of the carrier channel which flows into the outlet orifice. The carrier fluid contributes little to the movement of the dispersed fluid towards the outlet, until the dispersed fluid bulges out with sufficient volume to reach the high velocity flow (t=0.768). From here on it gets rapidly squeezed and accelerated through the outlet orifice, but under certain carrier flow rates, 30ml/h in this example, the trailing edge which contains two fluids, is pinched off. The emulsion-pinching effect observed in the enlarged flow-focus results in undesirable compound satellite emulsion(s) along with the main compound emulsion (t=0.256-0.384). For the cascaded COFUS junction the re-orientated carrier fluid flow aids the movement and shaping of the dispersed fluid towards the outlet at lower flow rates than its flow-focus counterpart. Therefore lower carrier flow rates can be used, which reduces the risk of unwanted emulsion-pinching.

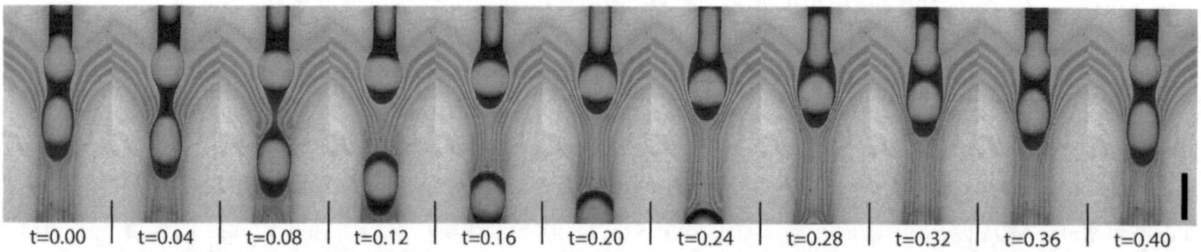

Figure 9: Footage at the second COFUS junction shows the production of compound emulsions which consists of a water inner fluid, TMPTA monomer outer fluid and a carrier of mineral oil. A small TMPTA satellite droplet is created behind the main droplet t=0.12. Blue streak lines are to aid carrier flow visualisation. Orientation of the carrier fluid aids the movement of dispersed fluid(s) towards the outlet. The curves create a flow-focusing orifice minima which aids breakoff mechanism when blocked by the dispersed fluid(s). Flow rates (ml/h); Inner(yellow) = 2 / Outer(red) = 2 / Carrier(blue/clear) = 10. Time in seconds. Scale bar is 2mm.

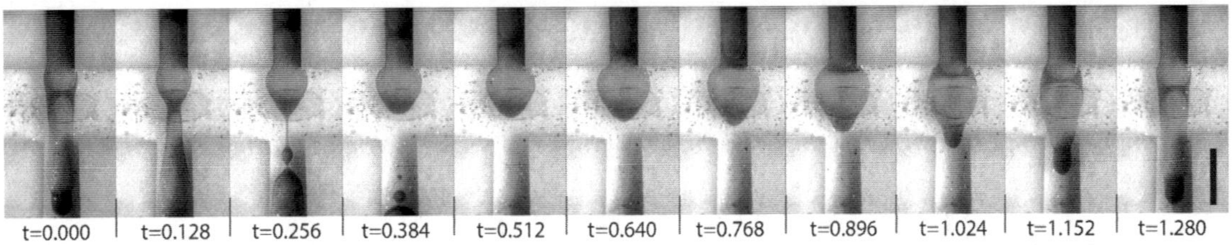

Figure 10: An enlarged flow-focus junction producing compound emulsions consisting of a water inner fluid, TMPTA monomer outer fluid and a carrier of mineral oil. The small particles are for velocimetry measurement. Compound satellite droplets can be observed to be formed at t=0.384 due to the rapid acceleration of the main emulsion once it reaches the high velocity flow around the sharp, lower corners. The high velocity flow pinches the rear of the breakoff to form the compound satellites (t=0.256-0.384). Flow rates (ml/h); Inner = 6 / Outer = 3 / Carrier = 30. Time in seconds. Scale bar is 2mm.

2.3 The Co-Flow-Focus in One Step (COFON) junction

An alternative design uses a one-step emulsion breakoff process with a junction called a co-flow-focus in one step or COFON (Figure 11). Here the two dispersed fluids are arranged coaxially using a slight offset between the fluid entrance nozzles to introduce one fluid inside another. The dispersed fluids are then broken off into compound emulsions by the carrier fluid flow within the co-flow-focus structure (Figure 12). Because the breakoff occurs over both the inner and outer fluids, compound satellites form from the thin strata of the breakoff process (Figure 13). The rupture of a coaxial strata can also result in water droplets within the outer layer which create voids within the polymerised shell and indicates why compound satellites should not be allowed to coalesce with the main breakoff. **(COFON Video)** captures the strata rupture event within the COFON junction.

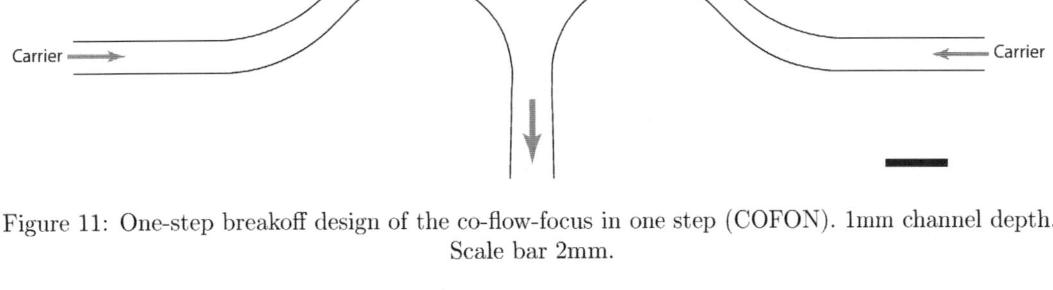

Figure 11: One-step breakoff design of the co-flow-focus in one step (COFON). 1mm channel depth. Scale bar 2mm.

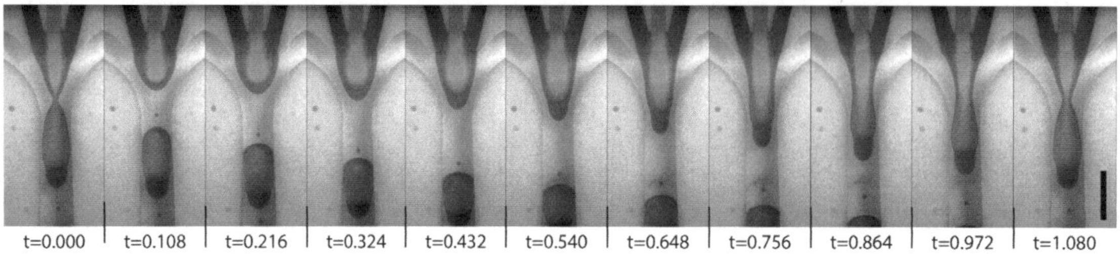

Figure 12: The COFON junction forming millimetre sized compound emulsions consisting of a water inner fluid, TMPTA monomer outer fluid and a carrier of mineral oil. A compound satellite droplet can be seen trailing the main breakoff, starting at t=0.108. Flow rates (ml/h); Inner=2 / Outer=4 / Carrier=14. Time in seconds. Scale bar is 2mm.

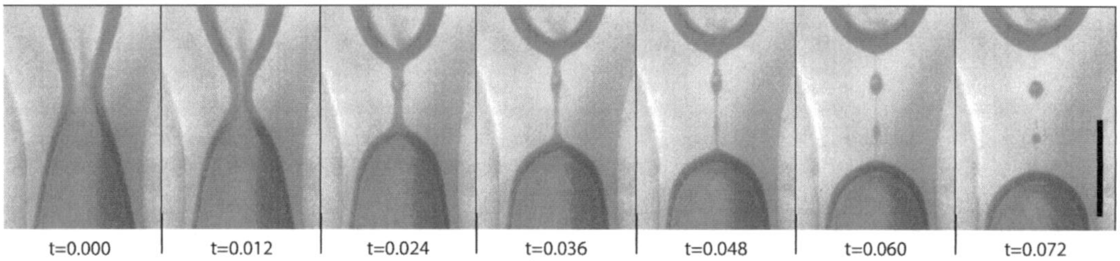

Figure 13: Close-up of the strata formed at the COFON junction and its subsequent rupture which forms compound satellites. Close inspection reveals a small water droplet formed by the rupture inside the TMPTA outer layer of the main breakoff towards the rear. Flow rates (ml/h); Inner=4 / Outer=3 / Carrier=14. Time in seconds. Scale bar 1mm.

3 Satellite droplet issues and amelioration strategies

If droplets of viscoelastic fluids are formed using an orifice diameter within a particular range, satellite droplets are likely to be formed[13]. Being a continuous system, satellite droplets flowing chaotically around opens up a can of worms. Discussed are two passive methods that can remove the presence of the satellites. In the two-step cascaded COFUS design the first junction produces a water in TMPTA emulsion. At the second junction the flow of mineral oil leads the constituent dispersed fluids to be separated into droplets at the TMPTA segment, which leads to a satellite containing only TMPTA. By allowing the satellite to coalesce once more with the main breakoff, the impact of chaotically flowing satellite droplets can be reduced. For coalescence to occur, the thin film of mineral oil separating the two bodies needs time to drain away. This is called the drainage time[14]. Coalescence was accomplished using a stepped diffuser (Figure 14) which reduces the velocity of the satellites to a greater extent than that of the main breakoff. The multi-step arrangement brings the bodies together for enough time and under enough pressure to coalesce (Figure 15). The one-step COFON design has extensive issues with satellite formation because they are compound satellites that are formed when the dispersed fluid, which contains two fluids, is dispersed as compound emulsions. If the compound satellites are allowed to coalesce with the main breakoff, a void will be created in the outer fluid of the main emulsion which cannot then be easily removed. One way around the satellite droplet issue is to remove them after formation, before they have come into contact with the main breakoff. It has been identified that the Dean-flow passive sorter (Figure 16) could be used to separate the satellites from the main emulsion, although it has not yet been tested for sorting larger emulsions; it has been shown to work effectively in other microfluidic systems[15].

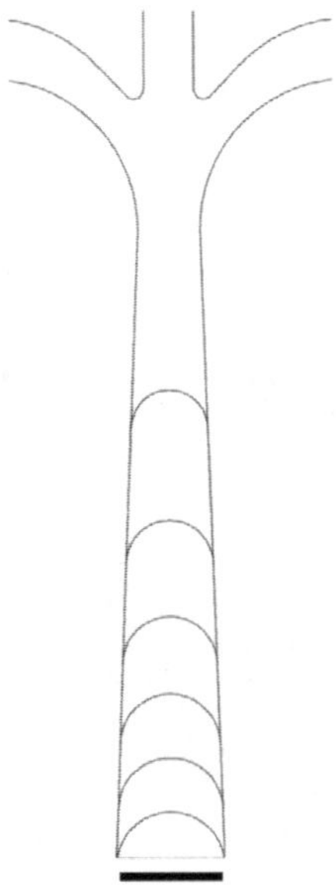

Figure 14: The stepped diffuser is joined directly after the second COFUS junction. Because the smaller satellite droplets have a lower velocity than the main breakoff, they eventually meet and press against each other until they coalesce. Each step is an increment of 0.2mm depth starting from 1mm at the junction. Distance between steps (semi-circle centre) is calculated by maintaining a consistent volume at each step, chosen to be equal to that of a 2.2mm diameter sphere. The steps ensure the confinement at the end of the diffuser does not limit the form of a droplet from being spherical. Scale is 2mm.

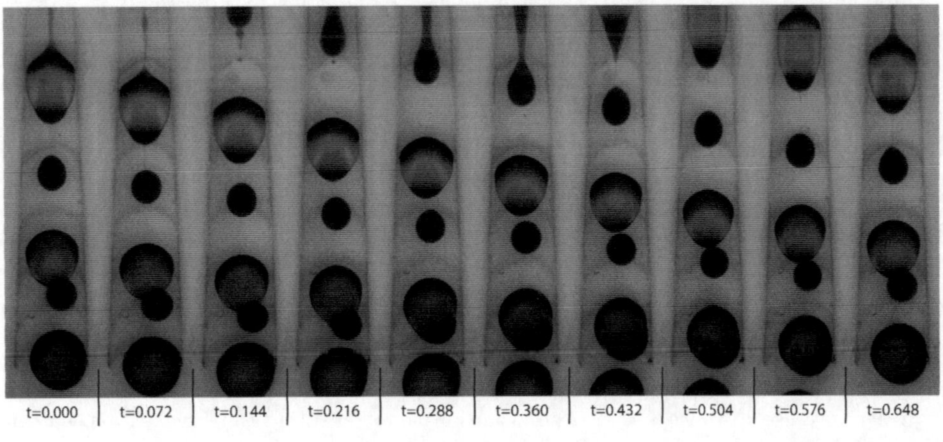

Figure 15: Footage of the stepped diffuser allowing satellites to coalesce with the main breakoff. Compound emulsions consist of a water inner fluid, TMPTA outer fluid and a carrier of mineral oil. Because a different clamping scheme was used in order to obtain unobstructed footage, flow leaked between the top of the side wall and the FEP film lid causing the emulsions to move to one side toward the end of the diffuser. For illustration of the coalescence, the carrier flow into the COFUS junction has been deliberately driven to exacerbate satellite formation. Satellites are comprised of TMPTA only. Flow rates (ml/h); Inner=9 / Outer=6 / Carrier=48. Time in seconds.

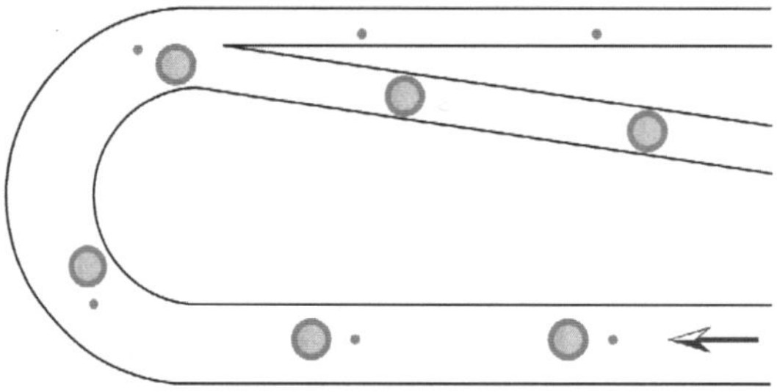

Figure 16: Passive microfluidic Dean flow sorter illustrating how a satellite droplet can be removed from the main breakoff using a curved channel.

4 Passive microfluidic methods for the concentric centring of compound emulsions

By far the most challenging task for the microfluidic method of shell manufacture is achieving micrometre-precision concentricity between compound emulsions in a predictable and cost effective way. Currently shells are manufactured in batches using flask-based chemistry, where agitation of the emulsions by propeller or by flask rotation during thermal polymerisation has been shown to significantly improve the concentricity[3][10]. However for IFE a batch process is unsuitable and a continuous method needs to be found. Di-electrophoresis centring techniques have indicated a possible strategy, however it is a relatively slow process, taking minutes[16]. NASA first observed that the fluids of a compound emulsion will become concentric if forced to oscillate at their resonant frequency[17]. Further on from the NASA work, a good model for the centring forces has been theorised by T. Norimatsu[18]. The model has focused on how the concentricity of the emulsion is affected by a step-response from a forced 'perturbation'. The model leads to the idea of using modulations of the channel itself i.e. by using 'passive compound emulsion centring by channel modulation'. Investigation of the influence that channel modulation has on the concentricity of a perturbated emulsion is an exciting area of research because the method could achieve high concentricity in a matter of seconds or even less.

One idea would be to use a single curve to create the step-response; however what was proposed was a design with multiple curved channel modulations. For an experiment sinusoidal modulations were added to a diffused photo-reactor's outer channel wall (Figure 17). The shells produced from the modulated design were compared to those produced by the control experiment where the diffused photo-reactor did not have modulations (Figure 18).

Once the compound emulsions are generated by the first two COFUS junctions and have any satellites recombined in the stepped diffuser, they reach what is known as the accelerator. The accelerator serves both to further increase distance between emulsions without affecting the breakoff mechanism of the junction(s) and to allow for fine tuning of the velocity of the emulsions over the channel modulations. The diffusing of the outlet channel creates an amplitude modulation envelope and allows a range of emulsion sizes to be perturbated. Photo-polymerisation of the TMPTA emulsions starts during modulation, using a row of eight 375nm low power ($\approx$0.2W) UV LEDs placed directly above the channel (Figure 22) and is fully polymerised when it passes two external high power 375nm ($\approx$7W) UV LEDs placed opposite each other and orthogonal to the outlet tube. Due to the size constraints of the chip the outlet channel could not be made flat, augment the modulation or apply sophisticated amplitude modulation. However the comparison of results displayed a clear improvement in the sphericity and concentricity of the polymerised shells when using channel modulation (Figure 19).

Figure 17: Modulated outlet diffuser.
Circle diameter is 50mm.

Figure 18: Unmodulated
outlet diffuser.

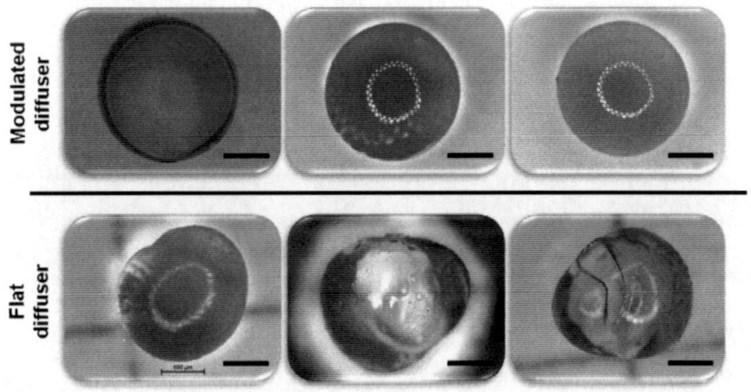

Figure 19: Photo-polymerised TMPTA shell comparison showing the effects of using a modulated or flat diffused photo-reactor. Intact shells retain the fluorescent inner water solution. Scale bar is 0.5mm.

5 Considerations for emulsion repeatability

The volume of each emulsion must be predictably repeatable to extremely tight tolerances in order to get a monodisperse population of shells. The fluid flow into the microfluidic junctions dictates the time period between the emulsions breakoff which relies heavily on cyclic pressure changes, caused when an emulsion blocks the output[19]. The fluid delivery system for the majority of microfluidic devices relies on the use of syringe pumps which are designed for dispensing volumetrically controlled fluid. Most syringe pumps use a motor connected to a lead screw to transform the rotary motion to a linear motion, thereby advancing the syringe's plunger. The motor used is most often a stepper type due to its high torque, low cost, wide speed range and quantised rotational steps, which allows for precise volumes to be dispensed over time. However the inherent 'cogging' due to the steps causes a non-continuous rotation which translates into a non-continuous linear motion advancing the syringe plunger. The effect of this non-continuous syringe driving is known to cause pressure fluctuation of the output fluid flow[20]. It is therefore reasonable to assume that a system that uses multiple syringe pumps, delivering at different flow rates, causes unsynchronised pressure fluctuations. These can disturb the breakoff mechanism, which is manifested in varying the time period between emulsion breakoff. The sensitivity of the breakoff process to changes in pressure and the use of multiple stepper-motor syringe pumps could explain the variation in emulsion volumes observed by Moynihan[11].

Journal of Physics: Conference Series **713** (2016) 012011 doi:10.1088/1742-6596/713/1/012011

5.1 Experimental method for testing emulsion repeatability

The time period between emulsion breakoff is used as the indicator for volume repeatability. From video footage the front of each successive emulsion passing an arbitrary line is timed to calculate the period between the occurrences for (n) instances. To compare the use of stepper and continuous syringe pump drives, two KdS 200 dual syringe pumps were employed. One of these had its stepper motor drive replaced with a precision ironless DC motor with a high reduction, low noise gearbox and rotary encoder. This type of motor does not suffer from any form of 'cogging' due to the ironless rotor. The syringes used were 1ml SGE gas-tight syringes which have a 4.61mm glass bore and PTFE plunger. One syringe was filled with mineral oil and the other with de-ionised water and together were fitted to the drive pump. The outlet from the syringes was attached to the first junction of the COFUS (Figure 8) in order to make W/O emulsions and was filmed at fifty frames per second. The DC motor has less speed range than the stepper motor and therefore dictates the flow rates that could be compared. To calculate the flow rate of the DC drive, the output frequency from the rotary encoder was recorded and averaged over the experiment. The average frequency could then be converted and used to set the equivalent flow rate on the stepper motor syringe pump drive.

5.2 Results from the emulsion repeatability experiment

Comparing the standard deviation of the breakoff period between the two drives shows that the DC drive has a lower variation than the stepper drive at all flow rates tested between 1.6 and 16.43 (ml/h). The equivalent rotational speed of the leadscrew is calculated to be 1.51 and 15.51 revolutions per minute for a 1.05735mm pitch lead driving a 1ml, 4.61mm diameter syringe. Table 1 gives the standard deviation and coefficient of variation of the breakoff period with the raw data presented in (Figure 20). For the lowest flow rate tested (1.6 ml/h) there is a notable peak in the breakoff period for the stepper motor drive at between instance 27 and 29 which is not observed from the DC drive. The sustained peak is possibly caused by the pump's internal feedback strategy which is likely to correct for missed steps and drive the syringe at an uneven velocity.

	(Standard deviation (s)) / [Coefficient of variation (%)]	
(Voltage (V)) [Equivalent flow rate (ml/h)]	DC drive	Stepper drive
(2) [1.6]	(0.160) [3.645]	(0.616) [13.507]
(4) [3.71]	(0.049) [1.907]	(0.080) [5.471]
(8) [7.91]	(0.015) [2.281]	(0.050) [5.513]
(16) [16.43]	(0.009) [2.179]	(0.018) [5.207]

Table 1: Variation in emulsion breakoff period for syringe pumps with a DC or stepper motor drive expressed as the standard deviation (s) and coefficient of variation (%)

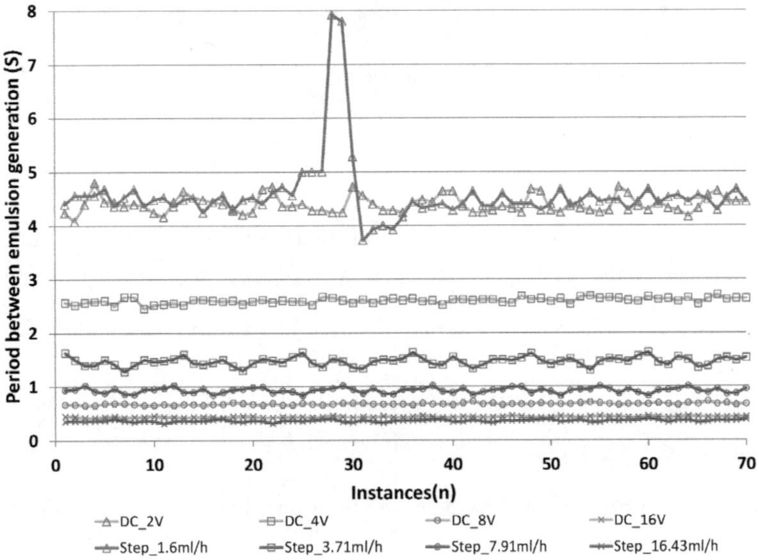

Figure 20: Variation in emulsion breakoff period for syringe pumps with a DC or stepper motor drive. Error bars are not included on the graph although are ±0.02s.

5.3 Conclusions from the emulsion repeatability experiments

Delivery of consistent pressure and flow rate to the emulsion generators is essential for forming consistent droplet breakoff timing which affects the monodispersity of shells. Although improvements can be made by the DC drive replacement, there is still undesirable variation of the breakoff period. The variation could be due to a number of factors such as the stick-slip characteristics of the syringes, the alignment of the dual syringe pumps pusher bar and lead screw accuracy among others. It is therefore reasonable to conclude that further improvements might be achieved by using other fluid delivery systems.

6 Cost of shell manufacture

Cost is of great concern, especially with the requirement of over one million shells per day. The carrier fluid used so far has been mineral oil. However it is relatively expensive and density matching is complex in contrast to using water as the carrier fluid to create a W/O/W emulsion. Using water as the carrier fluid increases the surface tension between it and the outer fluid (TMPTA). According to the most out of roundness (MOOR) law[3] a higher static sphericity can be achieved with increasing surface tension. In addition, diffusion is greatly reduced due to the reduced mutual miscibility between the fluids.

Formation of a W/O/W emulsion using the two-step cascaded COFUS requires hydrophillic walls at the second junction. The surface modification was achieved by coating a machined PMMA chip and lid in a hydrophillic material. This can be stacked on top of the first junction, machined into PTFE with a FEP film lid. The high contrast in static contact angle, with a water droplet (PTFE $> 100°$) and (Coated PMMA $< 5°$) enables W/O/W emulsions to be generated (Figure 21). The amount of usable shells that the microfluidic device can make in its lifetime versus the cost of the device must be understood as it will affect the final shell cost. Reliability might be improved if the base material or Lotus leaf patterning[21] was used to create hydrophillic and hydrophobic areas respectively.

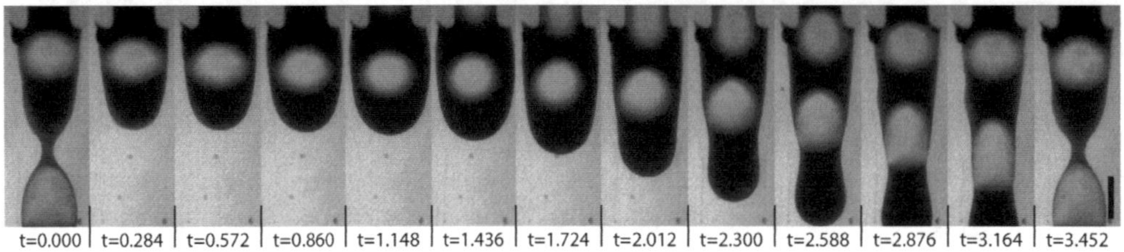

Figure 21: Millimetre sized W/O/W compound emulsions being formed at the second junction of the cascaded COFUS using a stacked chip of PTFE and a hydrophillic surface treated PMMA chip. Compound emulsions consist of a density matched water solution for the inner and carrier fluid and a TMPTA outer fluid. Flow rates (ml/h); Inner=1.5 / Outer=2 / Carrier=40. Time in seconds and scale bar is 1mm.

7 Conclusion

Microfluidics has shown a possible route to achieve the economical and rapid production of spherical polymer shells, although it is clear that one must control every part of the system to exceptionally tight tolerances in a holistic manner.

The novel planar geometries of the Co-Flow-Focus in One step (COFON) and cascaded Co-Flow-Focus (COFUS) generate compound emulsions in a one or two step breakoff method respectively. They have both shown an ability to produce 2mm+ size W/O/O double compound emulsions which satisfy the required dimensions for IFE. The COFUS has been further modified to generate W/O/W emulsions using PTFE for the first junction and coated PMMA for the second.

The formation of satellite droplets is observed using one and two step breakoff methods, the satellites themselves being comprised of either a single or a double fluid. If the satellite is of a single fluid, it can be recombined with the main breakoff by using a stepped diffuser; otherwise they could be removed using passive techniques such as the Dean flow sorter.

Use of channel modulations to perturbate and thereby aid concentric alignment of the emulsions is certainly an area of research to investigate further. A direct comparison between diffused photo-reactors with or without sinusoidal modulation showed an improvement in concentricity, sphericity and intactness when modulations are used.

A suitable fluid delivery system is one which is continuous and pulsation free, so as not to affect the emulsion breakoff process. Using an off the shelf syringe pump (KDS 200), variation in the time period between emulsions being generated was measured. Improvements were observed when the stepper motor drive was replaced with a geared down continuous brushed ironless DC motor. However the results of such modifications still fall short of requirements and other methods ought to be investigated.

By way of conclusion, controlled and cost-effective fusion energy must be achieved for the progression of mankind and the problem of the economic mass production of IFE shells might well be solved by using microfluidics, even though"nature is going to put up a fight" [22].

Figure 22: Science-selfie in front of the microfludic system that uses channel modulation to improve the concentric alignment of compound emulsions. Insert to the left shows photo-polymerisation of emulsions in a modulated and diffused photo-reactor. 27/02/2014.

References

[1] Siegel K M 1975 *J. Appl. Sci. Engin.* **1** 3–18

[2] Komeda O *et al.* 2013 *Sci rep* **3** 2561

[3] Cook R, Takagi M, McQuillan B and Stephens R 2000 The development of plastic mandrels for NIF targets Tech. rep. Lawrence Livermore National Laboratory

[4] Cordero M L, Gallaire F and Baroud C N 2011 *Phys. Fluids.* **23** 094111

[5] Sauret A and Shum H C 2012 *Int. J. Nonlinear. Sci. Numerical Simulation* **13** 351–362

[6] Utada S A, Chu L Y, Fernandez-Nieves A, Link D R, Holtze C and Weitz D A 2007 *Mat. Res. Soc. Bulletin* **32** 702–708

[7] Vladisavljevic G, Duncanson W J, Shum H O and Weitz D A 2012 *Langmuir* **28** 12948–12954

[8] Nakatsuka M, Tsubakihara H and Kubo U 1979 *Review of Laser Engineering* **2** 172–175

[9] Takagi M, Norimatsu T, Yamanaka T and Nakai S 1991 *J. Vac. Sci. Technol.* **9** 2145–2148

[10] Nagai K, Nakajima M, Norimatsu T, Izawa Y and Yamanaka T 2000 *J. Polym. Sci. Part A* **38** 3412–3418

[11] Moynihan M J 2013 *Double emulsion generation in the mass production of ICF targets using T-junctions* Ph.D. thesis University of Rochester

[12] Faith D, Horsfield C J and Nazarov W 2006 *J. Mat. Sci.* **38** 3973–3977

[13] Wagner C, Amarouchene Y, Bonn D and Eggers J 2005 *Phys. Rev. Lett.* **16** 164504

[14] Tan Y C, Ho Y L and Lee A P 2007 *Microfluidics and Nanofluidics* **3** 495–499

[15] Guan G, Wu L, Bhagat A A, Li Z, Chen P C, Chao S, Ong C J and Han J 2013 *Sci. Rep.* **3** 1475

[16] Bei Z 2010 *Dielectrophoresis based double emulsion droplet centering for concentric laser target foam shells* Ph.D. thesis University of Rochester

[17] Saffren M, Elleman D D and Rhim W K 1981 Normal modes of a compound drop *2nd Intenationa coll drop. bubbles* (NASA Jet propulsion laboratory) pp 7–14

[18] Norimatsu T, Izawa Y, Mima K and Gresho P M 1999 *Fusion Technology* **35** 147–156

[19] Baroud C, Gallaire F and Dangla R 2010 *Lab on a chip* **10** 2032–2045

[20] Li Z, Mak S Y, Sauret A and Shum H C 2014 *Lab on a chip* **14** 744–749

[21] Vorobyev A Y and Guo C 2015 *J. Appl. Phys.* 117

[22] Tolley M 2012

Acknowledgements

I would like to thank my sponsors STFC and Cardiff University for funding this PhD research *EP/L504749/1 (DTG2013).* Also to express one's gratitude to Martin Tolley, Eriketi Loizidou, Edwin Fox and my family for their technical and moral support and to Professor David Barrow for providing the facilities.